Understanding College Algebra

Through Inquiry and Logical Reasoning

Eric W. Kuennen
University of Wisconsin Oshkosh

Revised Second Edition. Copyright © 2021 by Eric W. Kuennen

This work is licensed under the Creative Commons Attribution-NonCommercial-NoDerivatives 4.0 International License. To view a copy of this license, visit http://creativecommons.org/licenses/by-nc-nd/4.0/ or send a letter to Creative Commons, PO Box 1866, Mountain View, CA 94042, USA.

Portions of this text have been adapted from the Big Ideas in Mathematics series of textbooks by John Beam, Jason Belnap, John Koker, Eric Kuennen, Amy Parrott, and Jennifer Szydlik, and are used with permission from the authors.

Cover image of a balance scale courtesy of the Metropolitan Museum of Art, through Wikimedia Commons.

This work was supported by a University of Wisconsin Oshkosh Faculty Development Grant.

First Printing: 2018. Second Edition 2019. Revised Second Edition 2021.

ISBN: 9798524639240

Eric W. Kuennen, Mathematics Department, University of Wisconsin Oshkosh
800 Algoma Blvd, Oshkosh, WI 54901

Contents

INTRODUCTION   1

Chapter One: The Foundations of Algebra

SECTION 1: FEATURES OF ALGEBRAIC THINKING   6
CLASS ACTIVITY 1: LET'S GET ACQUAINTED   6
READ AND STUDY 1: ALGEBRAIC THINKING   7
HOMEWORK SET 1   11

SECTION 2: INTERPRETING SYMBOLS, EXPRESSIONS AND EQUATIONS   13
CLASS ACTIVITY 2: SYMBOL SENSE   13
READ AND STUDY 2: DEFINING AND INTERPRETING SYMBOLS   14
HOMEWORK SET 2   18

SECTION 3: FUNCTIONS AS SEQUENCES OF OPERATIONS   20
CLASS ACTIVITY 3: FUNCTION MACHINES   20
READ AND STUDY 3: FUNCTION FORMULAS AND THE ORDER OF OPERATIONS   22
HOMEWORK SET 3   27

SECTION 4: PROPERTIES OF OPERATIONS   30
CLASS ACTIVITY 4: OPERATION   30
READ AND STUDY 4: THE LAWS OF ALGEBRA   32
HOMEWORK SET 4   34

SECTION 5: THE DISTRIBUTIVE LAW   36
CLASS ACTIVITY 5: CURSES, FOILED AGAIN!   36
READ AND STUDY 5: THE DISTRIBUTIVE LAW   38
PROBLEM SET 5   40

SECTION 6: ADDITIVE AND MULTIPLICATIVE INVERSES   42
CLASS ACTIVITY 6: PROVE IT!   42
READ AND STUDY 6: JUSTIFYING PROPERTIES   44
HOMEWORK SET 6   48

SECTION 7: USING ADDITIVE AND MULTIPLICATIVE INVERSES   49
CLASS ACTIVITY 7: CANCELLATION POLICY   49
READ AND STUDY 7: CANCELLATION FACTS AND MYTHS   51
HOMEWORK SET 7   56

SECTION 8: EXPONENTS — 58
- CLASS ACTIVITY 8: Exponentially Yours — 58
- READ AND STUDY 8: Exponent Definitions — 60
- HOMEWORK SET 8 — 63

SECTION 9: ROOTS OF NUMBERS — 65
- CLASS ACTIVITY 9: Let's Get Radical — 65
- READ AND STUDY 9: Roots of Numbers — 66
- HOMEWORK SET 9 — 71

SECTION 10: IRRATIONAL AND IMAGINARY NUMBERS — 73
- CLASS ACTIVITY 10A: Now You're Just Being Irrational! — 73
- CLASS ACTIVITY 10B: Me, Myself and i — 74
- READ AND STUDY 10: Irrational and Imaginary Numbers — 75
- HOMEWORK SET 10 — 80

SECTION 11: TESTING AND JUSTIFYING SIMPLIFICATIONS — 83
- CLASS ACTIVITY 11: Oversimplification — 83

Chapter Two: Equations

SECTION 12: EQUATIONS AND SOLUTIONS — 85
- CLASS ACTIVITY 12: Equality for All! — 85
- READ AND STUDY 12: Types of Equations — 86
- HOMEWORK SET 15 — 91

SECTION 13: EQUIVALENT EQUATIONS — 93
- CLASS ACTIVITY 13A: Follow the Law — 93
- CLASS ACTIVITY 13B: It's time again for "Equivalent or Not Equivalent!" — 94
- READ AND STUDY 13: Properties of Equality and Solving Equations — 95
- HOMEWORK SET 13 — 100

SECTION 14: GRAPHS AND TECHNIQUES FOR FINDING SOLUTIONS — 103
- CLASS ACTIVITY 14A: Let's Get Coordinated — 103
- CLASS ACTIVITY 14B: How Do You Solve an Equation Like Maria? — 104
- READ AND STUDY 14: Techniques for Solving Equations — 105
- HOMEWORK SET 14 — 110

SECTION 15: THE DISTANCE FORMULA — 111
- CLASS ACTIVITY 15: Go the Distance — 111
- READ AND STUDY 15: The Distance Formula — 112
- HOMEWORK SET 15 — 116

SECTION 16: FINDING EQUATIONS FOR GRAPHS	118
CLASS ACTIVITY 16: THROW ME A CURVE	118
READ AND STUDY 16: FINDING EQUATIONS FOR GRAPHS	119
HOMEWORK SET 16	121

SECTION 17: DEFINITION OF AN ELLIPSE	122
CLASS ACTIVITY 17: SQUASHED CIRCLES	122
READ AND STUDY 17: FINDING EQUATIONS FOR ELLIPSES	123
HOMEWORK SET 17	131

Chapter Three: Analyzing Functions

SECTION 18: FUNCTION DEFINITIONS	134
CLASS ACTIVITY 18A: DIRECTORY ASSISTANCE	134
CLASS ACTIVITY 18B: FUNCTION MACHINES REVISITED	135
READ AND STUDY 18: FUNCTION DEFINITIONS	136
HOMEWORK SET 18	140

SECTION 19: FUNCTIONAL THINKING	142
CLASS ACTIVITY 19A: BRIDGE TRUSSES	142
CLASS ACTIVITY 19B: GRAPHING RELATIONSHIPS	143
READ AND STUDY 19: FUNCTIONAL THINKING	144
HOMEWORK SET 19	147

SECTION 20: FUNCTION FORMS	149
CLASS ACTIVITY 20: COLONY DATA	149
READ AND STUDY 20: LINEAR, EXPONENTIAL AND QUADRATIC GROWTH	151
HOMEWORK SET 20	153

SECTION 21: LINEAR FUNCTION FORMS	155
CLASS ACTIVITY 21: CROSSING THE RIVER	155
READ AND STUDY 21: LINEAR FUNCTION FORMS	156
HOMEWORK SET 21	159

SECTION 22: QUADRATIC EXPRESSIONS	161
CLASS ACTIVITY 22: SQUARE DANCE	161
READ AND STUDY 22: AREA MODELS FOR QUADRATIC EXPRESSIONS	163
HOMEWORK SET 22	167

SECTION 23: QUADRATIC FUNCTIONS	168
CLASS ACTIVITY 23A: FORM FOLLOWS FUNCTION	168
CLASS ACTIVITY 23B: PARABOLAS AND QUADRATIC FUNCTIONS	169
READ AND STUDY 23: QUADRATIC FUNCTION FORMS	170
HOMEWORK SET 23	174

SECTION 24: TRANSFORMATIONS OF FUNCTIONS	**176**
CLASS ACTIVITY 24A: TRANSFORMATIONAL THINKING	176
CLASS ACTIVITY 24B: ON THE MOVE	177
READ AND STUDY 24: TRANSFORMATIONS OF FUNCTIONS	179
HOMEWORK SET 24	183
SECTION 25: POLYNOMIALS	**185**
CLASS ACTIVITY 25A: POPCORN BOXES	185
CLASS ACTIVITY 25B: FILL 'ER UP!	186
READ AND STUDY 25: THE FUNDAMENTAL THEOREM OF ALGEBRA	187
HOMEWORK SET 25	190

Chapter Four: Rational Functions, Logarithms, and Inverse Functions

SECTION 26: RATIONAL FUNCTIONS	**193**
CLASS ACTIVITY 26: LET'S BE RATIONAL ABOUT THIS	193
READ AND STUDY 26: RATIONAL FUNCTIONS	195
HOMEWORK SET 26	197
SECTION 27: EXPONENTIAL AND LOGARITHM FUNCTIONS	**200**
CLASS ACTIVITY 27A: PAPER TO THE MOON	200
CLASS ACTIVITY 27B: GRAPHS OF EXPONENTIAL AND LOG FUNCTIONS	201
READ AND STUDY 27: THE DEFINITION OF THE LOGARITHM FUNCTION	202
HOMEWORK SET 27	203
SECTION 28: THE NATURAL BASE e AND PROPERTIES OF LOGARITHMS	**206**
CLASS ACTIVITY 28A: PROPERTIES OF LOGARITHMS	206
CLASS ACTIVITY 28B: GROWING ALL THE TIME	207
READ AND STUDY 28: PROPERTIES OF LOGARITHMS AND THE NATURAL BASE e	208
HOMEWORK SET 28	211
SECTION 29: INVERSE FUNCTIONS	**214**
CLASS ACTIVITY 29A: BACK AND FORTH	214
CLASS ACTIVITY 29B: INVERSE FUNCTION MACHINES	215
READ AND STUDY 29: INVERSE FUNCTION DEFINITION AND NOTATION	216
HOMEWORK SET 29	221
SECTION 30: FINDING INVERSE FUNCTION FORMULAS	**223**
CLASS ACTIVITY 30: MORE INVERSE FUNCTION MACHINES	223
READ AND STUDY 30: FINDING INVERSE FUNCTION FORMULAS	224
HOMEWORK SET 30	231
SECTION 31: SOLVING EQUATIONS REVIEW	**233**
CLASS ACTIVITY 31: FINDING THE RIGHT NUMBER(S)	233

APPENDIX	236
REFERENCES	238
GLOSSARY	239

INTRODUCTION

Ours is not to reason why, ours is but to do or die.
Alfred, Lord Tennyson, Charge of the Light Brigade.

READ THIS! It will help you to understand this book.

This book is all about understanding algebra. We know this is not your first time studying algebra. We assume you have already had one or two years of algebra in high school, and perhaps even a refresher course in college. Most likely the emphasis in these previous algebra classes has been on memorizing procedures and acquiring skills: how to simplify different types of expressions, how to solve different types of equations, which formulas to use to solve which kinds of problems. You were shown **how** to do a particular math problem, and then practiced that procedure by doing many exercises similar to the example shown. But most likely the emphasis has not been on **why** these procedures work, and so you may not realize that algebra **makes sense** and is something that you yourself can figure out and explain why it works. Algebra skills are an important key to success in college, but we believe that conceptual understanding of algebra is the key to procedural fluency.

When am I going to need this? You are likely taking college algebra because it is a prerequisite for another course you need to take later on, such as calculus, statistics, chemistry, physics, business, economics, etc. Algebra is an extremely useful tool that is used in many fields. Often, college algebra texts will include many application problems involving all sorts of different phenomena: profit, revenue and cost; temperature, pressure and volume of gasses; exponential growth of bacteria or radioactive decay; position, velocity and acceleration of objects; linear regression of data, and so on.

While we do include some of these situations as context for some problems, the focus of this book is not exploring these applications and to see algebra in various contexts, what is often called "real world" problem solving. We leave that to your subsequent courses that use algebra.

Our reasons for this are intentional. If algebra is merely a tool for doing other things, then perhaps all we need is an instruction manual for how to use that tool. (But then you don't have a *mathematics* book, you'd have only a book that uses mathematics).

The problem with this is, without understanding the tool, that how-to manual has to give different instructions for every different application of the tool. And if there are lots of different situations in which students will be using that tool, the manual will need to be chock full of lots of procedures for each possible situation. In a college algebra course, for example, procedures for solving many different types of equations: linear equations, quadratic equations, cubic equations, equations with fractions, equations with square roots, equations with cube roots, equations with exponentials, equations with logarithms; not to mention all the various contexts: equations for chemistry, equations for physics, equations for economics.

But if you **understand** the tool, understand how it works, and why it works, then you can apply that tool to any situation. In a College Algebra course, if you understand what an equation **is**, and what it **means** to solve an equation, and understand the basic properties of equations, then you can apply this understanding to any kind of equation in any kind of context.

So our focus in this book is on the mathematics itself, that it, the logical reasoning that underlies the various techniques and procedures in algebra. Instead of "real world" problem solving, you might call our focus "math world" problem solving. We believe that by understanding the mathematics of algebra, the logic of it, the connections between ideas, and above all, that idea that algebra makes sense, that will provide you with the best preparation to later be able to apply this understanding to any other context.

This book aims to provide you with a very powerful tool, that of mathematical reasoning. We will provide you with opportunities and will expect you to reason logically, figure things out, and make arguments. These are skills that will help you in any future endeavor.

Putting the College into College Algebra. This course will not be a repeat of your high school algebra. We will be taking a more mature perspective. A college mathematics course, especially one that counts towards a general education requirement in mathematics, should introduce you to the big ideas of mathematical thinking. At the heart of mathematics is the notion of proof. Mathematics is not about getting the right answer; it's about figuring things out. Mathematics is about logical reasoning and being able to justify that what you claim is true. Often in school mathematics reasoning and proof has been relegated only to the subject of geometry, but all mathematics, not just geometry, is built upon logical reasoning and proof.

Algebra works because it makes sense. It is **not** something that needs to be memorized. Everything in algebra can be reasoned. You **can** figure it out. In this book, we will focus how all of the different properties and procedures and techniques in algebra follows logically from a few basic assumptions and definitions about how numbers work. You will see that algebra is really just arithmetic when one or more of the numbers involved is unspecified or unknown. So if you can do arithmetic, that is, adding and subtracting, multiplying and dividing, you can do algebra. It's just a matter of getting used to reasoning about numbers and expressions that represent numbers, as opposed to simply performing calculations by using a calculator or following a procedure.

Class Activities. Each section of this text starts with a Class Activity that is designed to engage you in actually *doing* mathematics in class. Doing math is not just calculating or mimicking a procedure that has just been shown to you. Actually doing math is figuring things out: investigating, making and testing conjectures, making arguments, and communicating your reasoning to others.

The Class Activity problems are intended to be worked on in small groups and are designed to bring up the big algebraic ideas to be discussed afterwards as a class. These are true problems that you are not expected to know how to solve right away. We will not be showing or telling you how to solve these problems, and although we usually will discuss these problems in the reading that follows, in general we will refrain from supplying you with answers and solutions to the class activity problems. It's your job in your small groups and as a class to arrive at an understanding of the problem and your solution that is deep enough so that you will be able to convince yourself and others that you are right.

> *One of the big misapprehensions about mathematics that we perpetrate in our classrooms is that the teacher always seems to know the answer to any problem that is discussed. This gives students the idea that there is a book somewhere with all the right answers to all of the interesting questions, and that teachers know those answers. And if one could get hold of the book, one would have everything settled. That's so unlike the true nature of mathematics.*
>
> *Leon Henkin*

This book is intended to be read. If you think about it, that's a rather strange thing to have to say. Of course, you may say, authors write books intending them to be read. But we know that often math textbooks do not end up being read by students, but instead are used by students as a reference to show them step by step how to solve various types of problems. In fact, many math textbooks are not designed to be read, and are presented as a "how-to" manual with examples for the to follow to be able to get the right answers to the homework problems.

We expect you to read the "read and study" sections slowly and carefully, with pencil in hand. We will often pose questions that we intend you to think about and answer before reading on. When we do work out some examples, we do so to be able to discuss the big ideas and illustrate our reasoning, not with the intention of providing you with models to copy. When reading worked examples, you should follow along mentally with us, doing your own calculations and confirming that what we have said makes sense to you.

> *Don't just read it; fight it! Ask your own questions, look for your own examples, discover your own proofs.*
>
> *Paul Halmos, I Want to Be a Mathematician*

Exercises vs Problems. To us, an exercise is something that you already know how to do. There is value doing exercises in order to get stronger or to get better at doing that thing. A problem is something that you do not know how to do. If you already know how to do it, it's not a problem, it's an exercise. Problems take figuring out. In the homework, you will find both exercises which are routine and intended to give you more practice thinking about the big concepts as well as problems are intended to be problematic. They will take time to explore, develop and make connections, and in some cases, extend your reasoning to develop new ideas.

The Back of the Book. In the back of the book you'll find a glossary with the important mathematical definitions for your reference when working on problems. We do not include "answers" to the homework exercises and problems in the back of the book. We know that people like to have the validation and external confirmation of seeing that their solution matches the textbook's, but in our experience having the "answers" readily available usually gets in the way of learning mathematics. Why struggle and persevere to figure something out and understand it when you can just look it up? Furthermore, mathematics is not about getting the right answer. Mathematics is about figuring things out, understanding relationships, and making arguments to prove you are right. Knowing that you are right without having to look to a book or someone else to tell you are right is the mark of true understanding.

But that doesn't mean that you are on your own. We will do our best in the Read and Study sections to discuss the big ideas, offer explanations, and show you some good examples of problem solving and making mathematical arguments. And you will have your classmates and instructor to work on problems and discuss ideas with.

Chapter One

The Foundations of Algebra

Class Activity 1: Let's Get Acquainted

The members of the Harmony Club get each meeting off to a harmonious start by shaking hands. It is a club rule, in fact, that every pair of members must shake hands. This used to be quick when the club was small, but now the membership has grown to 100 members and the greeting is starting to take some time. How many handshakes does the club perform now?

Read and Study 1: Algebraic Thinking

The only way to learn mathematics is to do mathematics.
                                                            Paul Halmos

The "Handshake Problem" in the class activity is one of our favorite algebra problems. Notice that this is a true problem: you weren't shown how to do it, you needed to figure it out. You may have considered some simple examples first before trying to tackle the case of 100 people. You may have drawn diagrams, made a table, and looked for patterns. And the case of 100 people is large enough so that in order to know you are right, you need to justify your method and explain why your solution makes sense. You really need to solve the general problem for any number of people to know that you have the right number for 100 people.

Within this one problem, we can see many important features of algebraic thinking, which we will highlight in this section. First, algebra is **generalized arithmetic**. The handshake problem is stated originally for 100 people. But a complete understanding of the problem and its solution requires that we consider the problem for *any* number of people. The problem is transformed from a specific question (how many handshakes will there be if there are 100 people?) to a general one (how many handshakes will there be if there are n people? At its core, the arithmetic required for the specific question (that is, the additions and the multiplications) is the same as the arithmetic for the general question.

Let's consider a much simpler example to explain what we mean about algebra being generalized arithmetic. Consider this question:

> "There are 4 people in my family. I want to buy enough pencils so that each person has 2 pencils. How many pencils do I need?"

This is an arithmetic problem, as it involves the basic arithmetic operations (addition and subtraction, multiplication and division). You could reason that I need $2 + 2 + 2 + 2 = 8$ pencils, or you could reason that I need $2 \cdot 4 = 8$ pencils. But now what if I ask the following:

> "There are 7 people in my family. I want to buy enough pencils so that each person has 2 pencils. How many pencils do I need?"

Again, you could reason that I need $2 + 2 + 2 + 2 + 2 + 2 + 2 = 14$ pencils, or you could reason that I need $2 \cdot 7 = 14$ pencils. Now what if I ask the following:

> "There are 5 people in my family. I want to buy enough pencils so that each person has 2 pencils. How many pencils do I need?"

At this point you are saying: "This is getting ridiculous. It's the same thing! Just multiply the number of people in your family by 2 and that will tell you how many pencils you need." Once you start thinking like that, now you are thinking **algebraically.** You are no longer just doing arithmetic on specific numbers (such as 2 or 7), you are doing arithmetic on generalized quantities, such as "the number of people in your family".

One way to generalize the problem would be by asking:

> "There are n people in my family. I want to buy enough pencils so that each person has 2 pencils. How many pencils do I need?",

and you could answer by saying that I need $2 \cdot n$ pencils. Note that this means exactly the same thing as "multiply the number of people in your family by 2 and that will tell you how many pencils you need", it's just more compact and efficient to use the variable n to represent the unspecified size of my family.

Another big idea in algebra is that algebra is about the **relationships between variables**. In the Handshake Problem, we have a variable which is the number of people in the club (which we could call n), and another variable which is the number of handshakes that take place (which we could call H). The problem is asking us to figure out the precise nature of the relationship between these two variables. Specifically, we want to know how many handshakes there are for any given number of people. In other words, we want a formula that tells us what to do to the number of people n to get the number of handshakes H.

There is one particular kind of relationship between variables that is especially important, the **functional relationship**, which is when each value of the first variable determines exactly one value for the second variable, and we can think of one of the variables as an **input**, and the other as an **output** that is uniquely determined by the input. In the Handshake Problem, there is a functional relationship between n, the number of people in the club, and H, the number of handshakes that take place. This is because for each number of people in the club, there is exactly one number of handshakes that take place. We can think of the number of people in the club as an input and the number of handshakes that take place as an output. For example, if I input that there are 5 people in the club, I get an output of 10 total handshakes. If I input that there are 10 people in the club, I get that there are 45 total handshakes.

Mathematics is all about exploring and understanding relationships, and we can represent these relationships in a variety of ways. In algebra, relationships between two variable quantities are typically represented in three forms: with tables, with equations, and with graphs, and a key component to being proficient in algebra is being able to use and make connections among these three representations.

A **table** shows the relationship in its most explicit and form: which specific values for the one variable corresponds to which values for the other variable(s). To illustrate, let's return to the simple problem we discussed earlier in this section.

"There are n people in my family. I want to buy enough pencils so that each person has 2 pencils. How many pencils do I need?",

This problem describes a relationship between two quantities. One quantity is the number of people in the family, and the other is the number of pencils required. This relationship is simple enough that it can be described easily in words, namely that the number of pencils needed is twice the number of people in the family.
On the right is a way this relationship could be represented in a table.

Number of people in the family	Number of Pencils Required
0	0
1	2
2	4
3	6
4	8
5	10

An **equation** shows a relationship in its most general form by writing the operations that are done any value for the one variable to calculate the corresponding value for the other variable. In our pencil problem, this relationship could be represented by the equation

$$P = 2 \cdot n,$$

where n represents the number of people in the family, and P represents the number of pencils required. This equation says in symbols what we said earlier in words, that the number of pencils required is twice the number of people in the family. In doing so, it describes how to find the value of P for any n, or conversely, how to find the value of n for any P. If I want to know the size of the family that requires 32 pencils, I can substitute the number 32 for the P, and I get the equation $32 = 2 \cdot n$, which I can then "solve" to find that n must be 16 people. Finding solutions to equations is a big part of algebra, and we will return to this topic in later sections.

A third way to represent relationships is graphs. In the pencil problem, we could represent the relationship between the number of people in the family and the number of pencils required by plotting ordered pairs of numbers on a Cartesian coordinate plane. The word "coordinate" refers to having numbers paired together, and the word "Cartesian" refers to Rene Descartes, a French mathematician in the 17th century who popularized this way of representing points on the plane with pairs of numbers. Here's a graph of the pencil problem, with the number of people in the family on the horizontal axis, and the number of pencils required on the vertical axis.

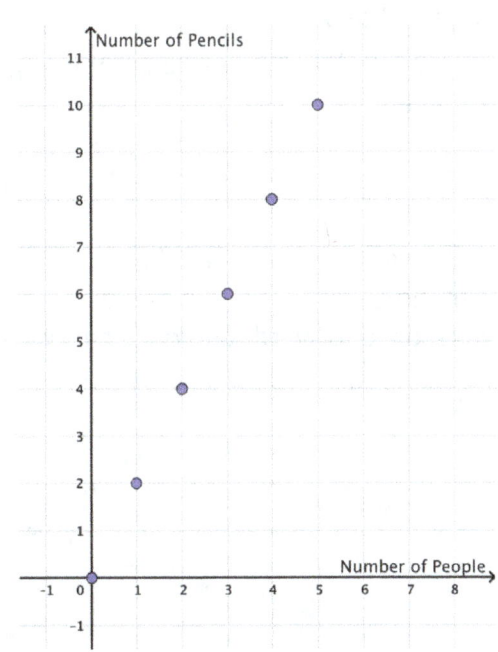

We know that our pencil example is a very simple one, but the big ideas of how tables, equations and graphs are used and interpreted are the same regardless how complicated the relationship between the variables might be.

Another great feature of the handshake problem is there are two different solution methods that people typically come up with. These two methods have a different structure, and lead to different general formulas, but they both make sense and of course yield the same solution, so it's interesting so see how the two solutions are algebraically equivalent. This brings up the important idea of different **algebraic forms**, that is, the value of writing the same quantities in different expressions. For example, when there are 5 people, there are 10 total handshakes. This information by itself is not very revealing. But if instead of just writing 10, we write 10 in an equivalent expression that reveals how it was computed, we can see the underlying structure of the problem. One way is to see this number 10 as 4 more than the number of handshakes when there are 4 people. Another way to see this is 3 + 2 + 1 + 0. Yet another way is to see it as half of 4 times 3.

The number of handshakes when there are 5 people is:

$$\text{(number of handshakes for 4 people)} + 4$$
$$4 + 3 + 2 + 1 + 0$$
$$\frac{1}{2}(5 \cdot 4)$$

All of the above forms are equivalent, in that they all equal 10. And each form makes sense in terms of the structure of the Handshake Problem. (In the homework, we will ask you to explain why each form makes sense. Moreover, each can be generalized to that the solution can be extended to 100 people, or for any number of people.

The fact that $4 + 3 + 2 + 1 + 0 = \frac{1}{2}(5 \cdot 4)$ is quite interesting. You already know that when you repeated add the same number together, that that is equivalent to multiplication. For example:

$$4 + 4 + 4 + 4 + 4 + 4 + 4 = 7 \cdot 4$$

Now in the expression $4 + 3 + 2 + 1 + 0$, we don't have the same number added together over and over. But we can create this situation with a very neat trick. Let's add this sequence up **twice**, once forwards and once backwards, and match up the numbers to get sums of 4.

$$\begin{array}{r} 4 + 3 + 2 + 1 + 0 \\ + \; 0 + 1 + 2 + 3 + 4 \\ \hline = \; 4 + 4 + 4 + 4 + 4 \end{array}$$

Since we have five 4's added together, the total is $5 \cdot 4$. But that's when you add up the sequence **twice**. Adding them up just once should give only **half** of that total. So that's why it makes sense that $4 + 3 + 2 + 1 + 0 = \frac{1}{2}(5 \cdot 4)$.

Now we called this a trick, but it's not magic. It worked because each successive number in our sum is changing by the same amount, so when we move over one matched pair, one of the numbers has gone up and the other has gone down by the same amount. So each pair adds to the same value. But we called it a trick because it's rather clever, and not an obvious thing to do.

There's an old legend about the famous German mathematician Friedrich Gauss (in fact, he's the one who first told the story) about this very method of adding up sequences. The story goes that while in elementary school, Gauss's teacher gave his students the task of adding up the numbers from 1 to 100. The teacher assumed that this task would keep the class busy for a long time (this was before calculators, but even with a calculator this would take a while) but it only took Gauss a minute to find the sum. Here's how he did it:

Add up the sequence twice, as follows:

$$\begin{array}{c} 1 + 2 + 3 + 4 + \cdots + 97 + 98 + 99 + 100 \\ + 100 + 99 + 98 + 97 + \cdots + 4 + 3 + 2 + 1 \\ \hline = 101 + 101 + 101 + 101 + \cdots + 101 + 101 + 101 + 101 \end{array}$$

So we get 101 added together repeatedly, in fact there are 100 of them, since that was how long the sequence was. So the total is $100 \cdot 101$, which is 10,100. But that's the sequence added up twice, so adding it up only once would be half that amount, which is 5,050.

Despite the impression you might have gotten in the past, there isn't just one way to solve a math problem. Typically there are a multitude of ways, and each can have its own value. Moreover, seeing the connections among different solution methods can have great mathematical power.

Homework Set 1

1) Another way to generalize the pencil problem from the Read and Study would be instead of varying the size of the family, we could vary the number of pencils per person. Answer the following: "There are 4 people in my family. I want to buy enough pencils so that each person has m pencils. How many pencils do I need?"

2) We could vary the size of the family and vary the number of pencils per person. Answer the following: "There are n people in my family. I want to buy enough pencils so that each person has m pencils. How many pencils do I need?"

3) Consider the Handshake Problem when there are 5 people. Make an argument that explains why it makes sense that the number of handshakes is:
    a) (the number of handshakes for 4 people) $+ 4$
    b) $4 + 3 + 2 + 1 + 0$
    c) $\frac{1}{2}(5 \cdot 4)$

4) Complete the following tables for the number of handshakes as a function of the number of people in the Harmony Club from the Class Activity.

Number of people in the club	Number of Handshakes
0	
1	
2	
3	
4	
5	
6	

Number of people in the club	Number of Handshakes
0	
10	
20	
30	
40	
50	
60	

5) Make an accurate graph on grid paper showing the data in the table above on the left, showing the number of handshake for 0-6 people in the club. Use a horizontal grid spacing of one person per grid line. Now make another graph showing the data in the table above on the right, showing the number of handshakes for 0-60 people in the club. Use a horizontal grid spacing of 10 people per grid line. Then answer the following questions:
    a) Is it appropriate to connect the dots in each of these graphs? Why or why not?
    b) Compare the shapes of the two graphs. Explain any differences you notice.

6) Suppose the Happy Couples Club consists of 100 married couples. That is, there are 200 people in the club, but each is married to one of the other members in the club. Before each meeting, every member of the club shakes hands with every other member who is not their spouse. How many handshakes take place? How many handshakes are there if the club has n total members?

7) Use Gauss's method to add up following sums.

  a) $1 + 2 + 3 + 4 + \cdots + 999 + 1000$
  b) $2 + 4 + 6 + 8 + \cdots + 100$
  c) $1 + 3 + 5 + 7 + \cdots + 99$
  d) $500 + 501 + 502 + 503 + \cdots + 599 + 600$

Class Activity 2: Symbol Sense

1) What numbers go in the boxes?

$$8 + 4 = \boxed{} + 5 = \boxed{}$$

2) Write a story problem that could result in writing the following equation:
$$60 = 3x + 4y.$$

Clearly state what number the variable x represents, and what number the variable y represents in your problem.

3) Shown is a picture of the border of a 6 by 6 square grid. Notice that there are 20 shaded border squares. Suppose you wanted a quick way to count the number of border squares in a square grid of size n. Your task is to write an algebraic expression for the number of border squares in an n by n grid.

Read and Study 2: Defining and Interpreting Symbols

Any impatient student of mathematics or science or engineering who is irked by having algebraic symbolism thrust upon him should try to get along without it for a week.
                                                                                                                    Eric Temple Bell

A common perception is that algebra is "math with variables." While it is true that we use many symbols in algebra, this is not the *point* of algebra. In fact, to a mathematician, algebra is the study of the structure of operations defined on sets. In this course, the sets we study are sets of numbers, such as the set of real numbers, and the operations we study are usually operations such as addition and multiplication. More about that later. For now, we want to remind you about some fundamental mathematical objects, symbols, and terms that you'll need to understand later ideas in this course.

First of all, we use symbols to represent numbers. In general, a **constant** is a fixed, unchanging number. For example, the symbol "7" represents this many objects:

■ ■ ■ ■ ■ ■ ■

So one type of number symbol is a **numeral** which is a symbol to represent a specific constant. We've gotten so used to the Arabic numerals $1, 2, 3, 4, 5, 6, 7, 8, 9$ that we often forget that they are symbols for numbers, and not the numbers themselves. The symbol "e" is typically used as a numeral, in the sense that "e" represents a specific number. So does the symbol "π". These are all examples of numerals.

While a numeral represents a specific constant number, a **parameter** represents an unspecified constant. Consider the standard meaning of the m in $y = mx + b$. Here m does not stand for one specific number (like 5), yet we think of it as *some* constant value, we just haven't specified it, at least not yet. So we would call the m and b parameters in the equation $y = mx + b$.

Suppose we do specify the constants m and b and have the equation $y = 2x + 3$. Now the symbols x and y can take on a whole lot of different values. The symbol x really represents any real number. So another type of number symbol is a **variable,** which represents an arbitrary value, or a changing quantity.

Moreover, the equation $y = 2x + 3$ expresses a functional relationship between the variables x and y. If $x = {}^-4$, then $y = {}^-5$. But if $x = 0.2$, then $y = 3.4$ The variable x stands for a changing quantity, and the value of y changes with x. In this case, x would be called the **independent variable** and y the **dependent variable**.

However, in mathematics, a letter symbol may not be referring to a number. Depending on the context, letters can be used to represent sets (usually capital letters $A, B, C, ...$), elements of sets (usually lower case letters $a, b, c, x, y, ...$), functions (usually lower case letters $f, g, h, ...$), or logical statements (often $P, Q, A, B, ...$).

In order to help with quickly interpreting symbols, people tend to use the same types of symbols for the same roles. For example, small letters near the end of the alphabet are often used to denote variables and small letters near the start of the alphabet often represent parameters. These are not hard-and-fast rules, but these conventions help us to not have to think about so much at once when looking at lots of letters.

Here is a glossary of typical uses for lower case letters in algebra:

a, b, c, d	unspecified constant
e	a specific constant (the natural exponent base), approx. 2.71828 ...
f, g	function
i	a specific constant (the square root of $^-1$).
n	integer-valued variable
t	real-valued variable (especially when it refers to a time)
x, y, x	real-valued variable

Some letters have two or more common uses. For example, in the setting of integer variables, after n has been used, the second choice is often m. But in the context of linear equations, m is used to denote the slope of the line. Also, if a third or fourth function letter is needed, h and k are common choices. However, h and k are also often used as parameters, such as the center of a circle being at the point (h, k).

And then we have symbols that stand for **operations**. For example, the 2 on x^2 represents a operation (called "squaring") that tells you to multiply x by itself. The pair of vertical bars in $|a|$ represents the operation of finding the absolute value of a number, which is how far the number is from 0 on the number line. On the other hand, + is a binary operation telling you to add *two* quantities. Also, there are **relational symbols** like = and < that specify a relationship between two quantities, sets, or objects.

A big part of doing algebra is representing variable quantities with symbols. Facility with defining and interpreting variables takes experience and merits further discussion. Consider the following scenario:

> At a certain school, there are three times as many students as teachers. Write an equation for this relationship, using S for the number of students and T for the number of teachers.

Did you have an equation written down? If not, go back and do it now. We really want you to think and work with us as you read.

Mark Driscoll, in his book *Fostering Algebraic Thinking*, discusses several "algebraic habits of mind". One key idea in identifying functional relationships between variables to solve problems is to focus on "**What changes? What stays the same?**" Clearly, in this scenario, the number of students can change. And the number of teachers can change. But the *relationship* between the number of students and teachers stays the same. There must always be three times as many students as teachers.

Another "Algebraic Habit of Mind" is **thinking about computations independently from the numbers used**. In our scenario we are being asked to write down how to compute the number of teachers given a number of students, or conversely, how to compute the number of students given a number of teachers, without using specific numbers of teachers or students. This is quite hard to do without first thinking about specific numbers. To write an equation that shows the relationship between the number of teachers and the number of students, we should first think about some specific numbers. For example, if there are 10 teachers in the school, there would be 30 students. If there are 40 teachers in the school, there would be 120 students. In general, we see that we could compute the number of students by multiplying the number of teachers by three. Or if we knew the number of students, we could find the number of teachers by dividing the number of students by three.

This process of first thinking about specific numbers, focusing on how they are related, and then generalizing those relationships and computations to be independent of the specific numbers is a key component of algebraic thinking.

So for your equation, did you write $3S = T$? In our experience, this is the equation that students most commonly write. But this equation is not correct. *Why not? Make an argument. What is a correct equation for the relationship? If the equation you wrote down at first turned out to be incorrect, why do you think this task turned out to be difficult?*

Let's talk a bit more about variables. In the equation above, the variable T doesn't represent teachers. We'll stay that again. The T doesn't represent teachers. Sure, we chose the letter T for the name of the variable because the word teachers starts with the letter T. But "teachers" is not a variable. A variable is a **quantity.** It is crucial to understanding algebra to realize that the variables in algebraic expressions are **numbers.** So T is a number. It is the **number** of teachers at the school. In the exercises, we will ask you to think more about the meaning of variables used in representing problems.

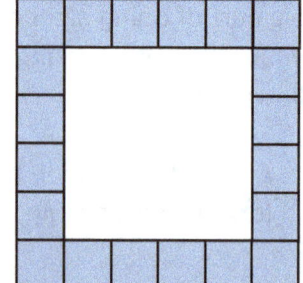

In the class activity, you considered the following problem. Shown is a picture of the border of a 6 by 6 square grid. Notice that there are 20 shaded border squares. Suppose you wanted a quick way to count the number of border squares in a square grid of size n. Your task is to write an algebraic expression for the number of border squares in an n by n grid.

Some good strategies to approach this problem are to look at several examples for various values of n, try to see the relationship between the size of the grid and the number of border squares, and generalize the computations involved. To really know why and be able to convince someone else that your formula is correct, you need to be able to explain makes it make sense based on some generalizable method for counting the squares.

Here is a list of some possible formulas for the number of border squares in an n by n square grid:

$$4 \cdot (n-1)$$
$$4 \cdot n - 4$$
$$4 \cdot (n-2) + 4$$
$$2n + 2(n-2)$$
$$n + 2(n-1) + (n-2)$$
$$n^2 - (n-2)^2$$

In mathematics, an **expression** is simply a combination of symbols. In real number algebra, an **algebraic expression** is an expression involving numbers (constants and variables) and operations (such as addition, subtraction, multiplication, division, powers and roots). In later sections, we will look more closely at symbols, and the properties of these operations. And in fact, we'll even look at different kinds of algebras where the variables don't represent numbers, and the operations are not the ones from arithmetic.

We will pause to note that students often refer mistakenly to all expressions as "equations". The expressions above are not equations, since they are not saying that one quantity equals another. An **equation** is a statement that says that two quantities or expressions have the same value. Simply put, an equation has an equal sign! However, we also saw in the class activity how important it is that we all interpret the meaning of an equals sign in the same way. In algebra, an equals sign doesn't mean "the answer is", but it means that the expressions on either side of the equals sign have the same value.

So did you come up with one of these expressions above? Could more than one of these be correct? Actually each of the above expressions are equivalent to each other. That means if you substitute the same number n into each formula, they will all give you the same value. So they are all "correct". Moreover, each one of these formulas represents a different valid way for computing the number of border squares. In the exercises, we will ask you to figure out exactly how each of these formulas counts the number of border squares.

You have probably have already had a lot of practice "simplifying" algebraic expressions. Which one of the above formulas is "simplified"? Well, it depends on your perspective. Each formula in its own way shows as clearly as possible the method used for counting the squares, so each of these forms has great value. What form you want to write an expression in depends on what information you want to make clear to the reader. The ability to write expressions in different equivalent forms and to appreciate the meaning of these different forms is an important algebraic skill.

So why is it that so many math texts and teachers typically show you only one solution method and one correct answer? Perhaps they have fallen into the misconception that math is about getting the right answer. So once they show you a simple efficient way of getting that answer, why discuss another? But math isn't about getting the right answers. (If all that is important is the right answer, then just look it up (like in the back of the book!). All these problems have been done before. (Or you can just scan the problem into your phone and have your app do it for you.)

No, math isn't about the right answer. Math is about figuring things out. It's about reasoning, exploring patterns, understanding relationships, and arguing that you are right. So if you have a problem that someone has already solved one way, but you see a new pattern, a different way of seeing it, new connection that can be made, then you are doing some great mathematics.

Homework Set 2

1) In the Read and Study, we asked you about the following scenario:

   > At a certain school, there are three times as many students as teachers. Write an equation for this relationship, using S for the number of students and T for the number of teachers.

   a) Explain why the equation $3S = T$ does not represent the relationship between the number of teachers and number of students at this school.
   b) What is a correct equation for the relationship? Argue that you are correct.

2) Here is an expression for the cost of 7 shirts and 3 hats: $C = 7x + 3y$. What precisely does the x represent? What type of symbol is the x? The y? The C?

3) Here is an expression for the cost in (dollars) of purchasing n pineapples and m papayas:

   $$5n + 2m$$

   a) What is the meaning of the number 5 in the expression?
   b) Which costs more: a pineapple, or a papaya? How can you tell?
   c) Which fruit is there more of: pineapples, or papayas? How can you tell?
   d) Is it mathematically acceptable that m and n be the same number even though they are symbolized by different letters? Explain.

4) (Note: This problem is different from the previous one. Read it carefully!)
   Here is an expression for the cost in (dollars) of purchasing pineapples and papayas if pineapples cost n dollars each and papaya cost m dollars each:

   $$5n + 2m.$$

   a) What is the meaning of the number 5 in the expression? Explain.
   b) Which costs more: a pineapple, or a papaya? How can you tell?
   c) Which fruit is there more of: pineapples, or papayas? How can you tell?
   d) Is it mathematically acceptable that m and n be the same number even though they are symbolized by different letters? Explain.

5) In the Class Activity, we discussed the "Square Border" problem and gave several different formulas as solutions to the number of border squares in an n by n grid. Justify each formula by explaining how it counts the number of border squares, based on the structure of square border.

   a) $4 \cdot (n-1)$

   b) $4 \cdot n - 4$

   c) $4 \cdot (n-2) + 4$

   d) $2n + 2(n-2)$

   e) $n + 2(n-1) + (n-2)$

   f) $n^2 - (n-2)^2$

Class Activity 3: Function Machines

A **function** can be thought of as a process that takes an input and ultimately returns a unique output. The notation $y = f(x)$ means that the values of the variable y are the outputs of the function named f with inputs x.

A powerful way of visualizing a function is with a "machine diagram" that shows a sequence of operations. For each machine diagram below:
  a. Find the intermediate outputs and the final output if you input the number 4.
  b. Find the intermediate outputs and the final output if you input the number $^-2$.
  c. Find the intermediate outputs and the final output if you input the number x.

1.

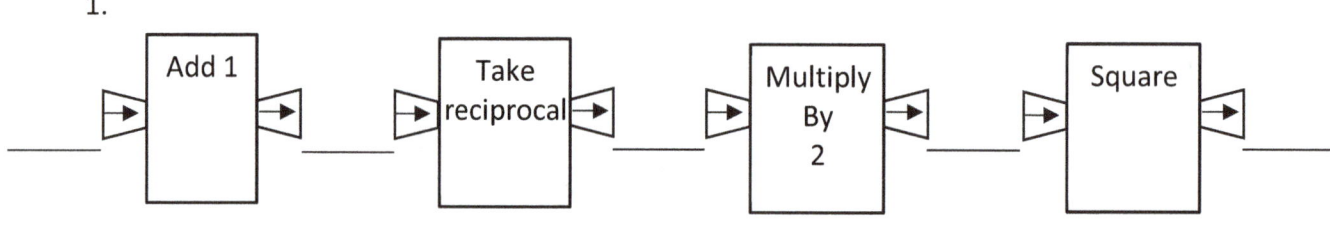

2.

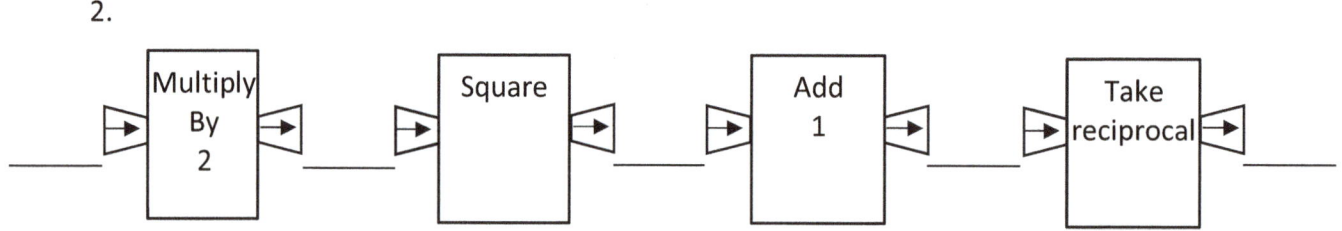

3.

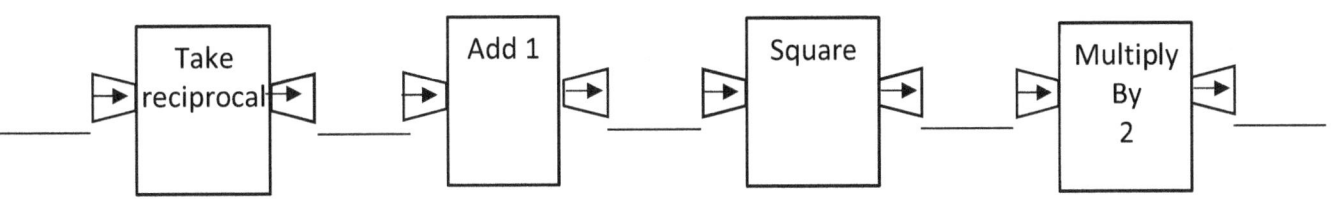

4.

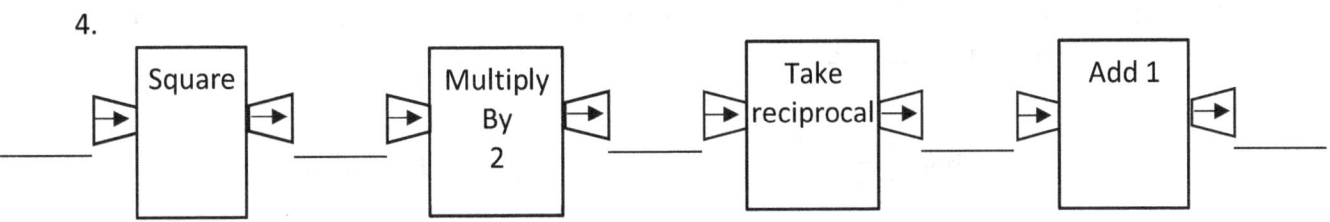

Read and Study 3: Function Formulas and the Order of Operations

Civilization advances by extending the number of important operations which we can perform without thinking about them.

Alfred North Whitehead

The concept of the function is one of the most important concepts in this course. We will start by understanding the definition:

> A **function** is a relationship between two sets in which each member of one set (called the **domain**) corresponds to exactly one member of another set (called the **range**).

In the Handshake problem from the first section, there is a functional relationship between the number of members of the club and the number of handshakes that take place. For each number of members, there is exactly one number of required handshakes. If we let the variable n represent the number of club members, let the variable H represent the number of handshakes required, and let the function f represent the rule that assigns each number of members to a number of handshakes, then we can say that $H = f(n)$.

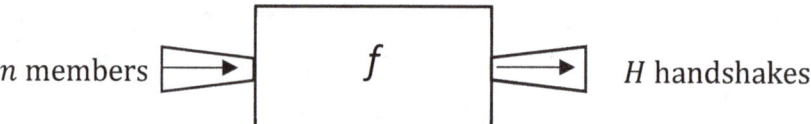

A function is a specific type of relationship between two variables. The variable whose values are elements of the domain is called the **independent variable** for the function, and the variable whose values are elements of the range is called the **dependent variable** for the function. *Why does it makes sense to call the domain variable independent and the range variable dependent?* In standard notation, x is used as an independent variable, and y as the dependent variable. Using these variables, a simple machine diagram for a function is as shown.

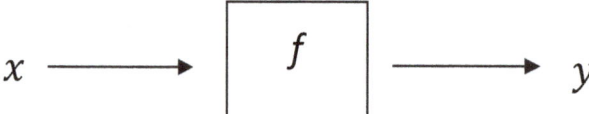

The idea represented by this diagram is more commonly written as an equation $y = f(x)$. Here, the symbol f is the name of the function and the notation $f(x)$ means that the variable x is the input into the function called f.

> **Function Notation**
>
> $y = f(x)$ means that y is the output when x is the input into the function called f.

The function notation $y = f(x)$ uses parentheses to designate the input into the function. Note that parentheses are also often used to denote multiplication, such as in the following expressions:

$$3(x - 1)$$
$$(x - 2)(x + 7)$$
$$a(^-b)$$

However, with the function notation $y = f(x)$, the "f" and the "x" are NOT being multiplied together. Moreover, multiplication wouldn't make sense in this case. Multiplication is an operation on two numbers. And while the x represents a number, the "f" does not represent a number. The f is the name of the function, which is a process, or a rule that relates the input x to the output y. We give each function a name because it's important to be able to think about the properties that that function has, and be able to those properties to those of other functions.

Note that in all of the expressions using parentheses listed above, the expression in front of the parentheses denotes a number: "3", "$(x - 2)$" and "a" all are representing numbers that are to be multiplied. When you think about it, using parentheses to indicate multiplication is just a case of people being lazy and leaving out the multiplication symbol. What is really meant by these expressions is:

$$3 \cdot (x - 1)$$
$$(x - 2) \cdot (x + 7)$$
$$a \cdot (^-b)$$

You may ask how we know that in $f(x)$ the "f" represents a function, while in $a(^-b)$, the "a" represents a number. Well, sometimes you just have to know from the context. But there are conventions in the choice of what letters we use. As we mentioned in the previous section, typically the letters f, and g, are "reserved" for denoting functions, and not used to denote numbers.

As we saw in the Class Activity, a powerful way to think of a function f is as a process that takes an input (from its domain) and ultimately returns a unique output (in the range). Often the process that goes on inside the function is a sequence of arithmetical operations where each operation can be thought of as its own sub-function or "machine". Writing out a function machine diagram showing each operation can be rather tedious. The good news is that there is a much more compact way to represent the exact same information, namely an algebraic expression for the output when the input is a variable x.

For example, the first function in the class activity could be written using function notation and an algebraic expression for the output:

$$f(x) = \left(\frac{2}{x + 1}\right)^2$$

which means that when the input into the function is the number x, the output of the function would be the number $\left(\frac{2}{x+1}\right)^2$. We can still think of this as a function machine, where there is only one compartment, and inside that compartment is a more complicated sequence of steps, which we can describe as $\left(\frac{2}{x+1}\right)^2$.

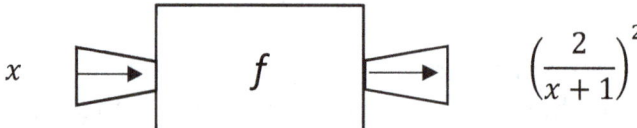

The other three functions in the class activity can be represented by the following three expressions (respectively):

$$\frac{1}{(2x)^2+1} \qquad 2\left(\frac{1}{x}+1\right)^2 \qquad \frac{1}{2x^2}+1$$

These algebraic expressions contain all the same information that the function machine diagrams contain, including the order of the sequence of operations to be performed. The key is, however, to be able to correctly read and interpret the algebraic expression with the right order of the operations, and that is the main point of this section.

In a function machine diagram, it is clear the precise order of the sequence of operations to be done. When writing this calculation as an algebraic expression, we need to be sure that the order of the operations is clear. You'll find that parentheses are very helpful in making it clear as to which operations, and we recommend that you use parentheses liberally whenever they help to make the meaning of the expression clear. But in the absence of parentheses, we (meaning the mathematics community: mathematicians, scientists, engineers, teachers, students, and so on), have adopted the following "Order of Operations" convention.

Order of Operations Convention

Evaluate an expression involving real numbers in the following order:

1. Evaluate any expressions inside parentheses or other grouping symbols

2. Evaluate exponents (powers) and roots

3. Evaluate multiplication and division as read from left to right

4. Evaluate addition and subtraction as read from left to right

For example, in expression $2 + 3 \cdot 4^2$, the exponent should be computed first, then the multiplication, then the addition, as follows:

$$2 + 3 \cdot 4^2$$
$$= 2 + 3 \cdot 16$$
$$= 2 + 48$$
$$= 50.$$

The expression $^-3^2$ is particularly problematic. While the negative sign really ought to be considered part of the numeral for the number 'negative three', under the order of operations agreement, this expression is usually interpreted with the exponent taking precedence, so that $^-3^2 = {^-9}$. If you want to write down that you are squaring the number negative three, this would be written as $(^-3)^2 = 9$. Our recommendation is that if there is any chance of misinterpretation, parentheses should be used for clarity: $^-(3^2) = {^-9}$, while $(^-3)^2 = 9$.

It's likely in school you've learned one or more mnemonic devices for remembering the order of operations, such as the acronym **PEMDAS** for "**P**arentheses, **E**xponents, **M**ultiplication, **D**ivision, **A**ddition, **S**ubtraction", or the phrase "**P**retty **P**lease **M**y **D**ear **A**unt **S**ally" for "**P**arentheses, **P**owers, **M**ultiplication, **D**ivision, **A**ddition, **S**ubtraction". However, these may be misleading if you are not aware that multiplication and division are of equal precedence, as are addition and subtraction. For example, if you were to interpret these to mean "addition first, subtraction afterward", you might incorrectly evaluate the expression

$$10 - 3 + 2$$

as being equal to 5, where the correct value under the order of operations convention is 9. The reason why Addition and Subtraction are given equal precedence is that subtraction is defined as addition of a negative number. In this case, the expression above is equivalent to the sum of 10, $^-3$ and 2:

$$10 + (^-3) + 2$$

so it has the value of 9. Similarly, division is defined as multiplying by a reciprocal, hence multiplication and division are given equal precedence.

Another drawback of the acronym PEMDAS is that the 'P' for Parentheses really represents all grouping symbols, and not all grouping symbols are as obvious as parentheses and brackets. For example, with roots, the extending the top horizontal line of the surd symbol $\sqrt{}$ is a really grouping symbol. For example, take a minute and determine what you think the value for each of the following three expressions should be:

$$\sqrt{9} + 16$$

$$\sqrt{9 + 16}$$

$$\sqrt{9} + \sqrt{16}.$$

In the first, the square root symbol only applies to the nine. Using the order of operations, we evaluate the root first, then do the addition:

$$\sqrt{9} + 16$$
$$= 3 + 16$$
$$= 19.$$

In the second, the square root symbol applies to the $9 + 16$, so it works as a grouping symbol. Since grouping symbols take precedence in the order of operations, the sum $9 + 16$ is to be computed first.

$$\sqrt{9 + 16}$$
$$= \sqrt{25}$$
$$= 5.$$

If one does not realize that the extended top bar of the square root symbol acts as a grouping symbol, then one may mistakenly think that the roots should be computed first before the sum, instead of the sum being computed before the root. If one in fact intends the roots to be taken first, then it needs be written as in the last expression:

$$\sqrt{9} + \sqrt{16}$$
$$= 3 + 4$$
$$= 7.$$

Similarly, an extended fraction bar acts a grouping symbol. For example, take a minute and determine what you think the value for each of the following three expressions should be:

$$\frac{1}{3} + 2$$

$$\frac{1}{3+2}$$

$$\frac{1+2}{3}$$

$$\frac{1}{3}+\frac{1}{2}$$

These are equivalent to $\frac{7}{3}, \frac{1}{5}, \frac{3}{3}$, and $\frac{5}{6}$ respectively. In the second and third expression, the extended fraction bar acts as a grouping symbol, so in each of those cases the additions should be calculated first.

The **absolute value** operation, which means to find the distance the number is away from zero on a number line, is symbolized by putting the number inside a pair of vertical bars. These vertical bars also make a grouping symbol, which means that the expressions inside the absolute value bars should be evaluated first. For example, the following four expressions all have different values. Take a minute and determine what the value should be for each.

$$|4 - 3 + {}^-5|$$

$$4 - |3 + {}^-5|$$

$$4 - 3 + |{}^-5|$$

$$4 - 3 - |5|$$

To emphasize that addition and subtraction have the same precedence (and multiplication and division have the same precedence), as well as the fact that there are many other grouping symbols besides parentheses, we suggest the Order of Operations Convention might be better represented by the acronym **GEMA**, for **G**rouping, **E**xponents, **M**ultiplication, **A**ddition.

Homework Set 3

1) Find the value of $f(4)$, if the function $f(x)$ has the given formula:

   a) $f(x) = \frac{x+8}{x}$
   b) $f(x) = x + \frac{8}{x}$
   c) $f(x) = \frac{x}{x} + 8$

2) Find the value of $f(4)$, if the function $f(x)$ has the given formula:
   a) $f(x) = \frac{12}{x} + 2$
   b) $f(x) = \frac{12}{x+2}$

3) Find the value of $f(4)$, if the function $f(x)$ has the given formula:
   a) $f(x) = \sqrt{x+5}$
   b) $f(x) = \sqrt{x} + 5$
   c) $f(x) = \sqrt{x} + \sqrt{5}$

4) Find the value of $f(3)$, if the function $f(x)$ has the given formula:
   a) $f(x) = (x-6)^3$
   b) $f(x) = x - 6^3$
   c) $f(x) = x^3 - 6$
   d) $f(x) = x^3 - 6^3$

5) Find the value of $f(^-1)$, if the function $f(x)$ has the given formula:
   a) $f(x) = {}^-3x^2$
   b) $f(x) = (^-3x)^2$
   c) $f(x) = {}^-(3x)^2$

6) Find the value of $f(^-1)$, if the function $f(x)$ has the given formula:
   a) $f(x) = 2x^3 - 5x$
   b) $f(x) = 5x - 2x^3$

7) Find the value of $f(^-3)$, if the function $f(x)$ has the given formula:
   a) $f(x) = \frac{9+x}{x}$
   b) $f(x) = \frac{9}{x} + x$

8) Find the value of $f(^-3)$, if the function $f(x)$ has the given formula:
   a) $f(x) = \sqrt{4+x^2}$
   b) $f(x) = \sqrt{4} + x^2$
   c) $f(x) = 4 + x$

9) Find the value of $f(^-3)$, if the function $f(x)$ has the given formula:
   a) $f(x) = 2(x - x^2)$
   b) $f(x) = 2(x - x)^2$
   c) $f(x) = (2x - x)^2$

10) For each function formula, make a machine diagram showing the sequence of operations done to x. Then find the value of $f(1)$ and $f\left(\frac{1}{4}\right)$.

a) $f(x) = \sqrt{\frac{1}{x} + 3}$

b) $f(x) = \frac{1}{\sqrt{x}+3}$

c) $f(x) = 2x^3 - 5$

d) $f(x) = (2x - 5)^3$

e) $f(x) = {}^-4 \cdot \left(\frac{1}{3x} + 2\right)$

f) $f(x) = |x - 3| + 3$

g) $f(x) = |x + 3| - 3$

Class Activity 4: Operation

OPERATION

1. Look up the definitions for Commutative, Associative, Identity Element, and Inverses in the Read and Study Section that follows this class activity. Using these definitions, fill in the table below to indicate whether each operation has the listed property on the given set.

	Addition (on the real numbers)	Subtraction (on the real numbers)	Multiplication (on the real numbers)	Division (on non-zero real numbers)
Commutative? Give an example or counterexample	Yes. Example: $3 + 4 = 4 + 3$			
Associative? Give an example or counterexample	Yes. Example: $(2 + 4) + 4 = 2 + (4 + 4)$			
Identity Element? If yes, what is it? Give an example or counterexample	Yes. It is 0. Example: $3 + 0 = 3$ and $0 + 3 = 3$.			
Inverses? If so, describe them. Give an example or counterexample	Yes. The inverse of a is $-a$. Example: The inverse of 4 is -4, because $4 + (-4) = 0$ and $(-4) + 4 = 0$.			

2. Operations can also be defined on sets of things that aren't numbers. The following table defines an operation # on the set of letters {A, B, C, D}. For example A # D = B.

#	A	B	C	D
A	A	D	A	B
B	D	D	B	C
C	A	B	C	D
D	C	C	D	B

a) Is this operation commutative? What did you have to check?

b) What would you have to check to see whether this operation is associative?

c) Does this operation have an identity element? If so, what is it?

d) Which elements have inverses?

Read and Study 4: The Laws of Algebra

A definition is a sack of flour compressed into a thimble.
                                                                    Remy De Gourmont

A (binary) **operation** on a set associates each ordered pair of elements in that set with exactly one element in that set. For example, addition is an operation on the set of real numbers since each pair of real numbers, when added, gives you exactly one number as their sum.

An operation is said to be **commutative** if the order in which the two elements are operated together does not change the result. For example, addition (on the set of real numbers) is commutative, because
$$a + b = b + a$$
for every choice of numbers a and b.

An operation is said to be **associative** if, when operating twice in succession, the choice of which of the two operations is performed first does not change the result. For example, addition (on the set of real numbers) is associative, because

$$(a + b) + c = a + (b + c)$$

for every choice of numbers a, b, and c. (Note: the parentheses, according the order of operations convention, mean that the a and b are added together first in the expression on the left, and that b and c are added together first in the expression on the right. Addition being associative means that either way, you get the same result, hence the equality).

An **identity element** for an operation is an element that when operated with any other element always results in that other element. For example, 0 is the identity element for addition on the set of real numbers, because
$$0 + a = a, \text{ and } a + 0 = a,$$
for every number a.

Suppose an operation has an identity. Then a pair of elements are **inverses** under that operation if when operated the result is the identity element. For example, ⁻4 and 4 are inverses with respect to addition, since they add to 0, which is the identity for addition. In general, a and ⁻a are inverses under addition, since

$$a + (^-a) = 0 \text{ and } (^-a) + a = 0.$$

Why make a big deal about these properties? You already know how to add, subtract, multiply and divide, right? We are making a big deal about the properties since these are the same properties of algebra. If you can add, subtract multiply and divide, you can do algebra. Really! Moreover, if you understand these properties, you can *understand* algebra!

The Laws of Algebra

Definition of Addition

Addition is defined with the following properties:

- **Commutativity:** For all real numbers a and b, $a + b = b + a$.
- **Associativity:** For all real numbers a, b and c, $a + (b + c) = (a + b) + c$.
- **Identity:** There is a real number '0' so that for all real numbers a, $a + 0 = a$.
- **Inverse:** For all real numbers a, there is a unique real number ^-a such that $a + {^-a} = 0$.

Definition of Subtraction

Subtraction is defined as adding the inverse:

- For all real numbers a and b, $a - b = a + (^-b)$.

Definition of Multiplication

Multiplication is defined with the following properties:

- **Commutativity:** For all real numbers a and b, $a \cdot b = b \cdot a$.
- **Associativity:** For all real numbers a, b and c, $a \cdot (b \cdot c) = (a \cdot b) \cdot c$.
- **Identity:** There is a real number '1' so that for all real numbers a, $a \cdot 1 = a$.
- **Inverse:** For all real numbers $a \neq 0$, there is a unique real number $\frac{1}{a}$ such that $a \cdot \frac{1}{a} = 1$

Definition of Division

Division is defined as multiplying by the inverse:

- For all real numbers a and $b \neq 0$, $a \div b = \frac{a}{b} = a \cdot \frac{1}{b}$

The Distributive Law

For all real numbers $a, b,$ and c:

$$a \cdot (b + c) = (a \cdot b) + (a \cdot c)$$

On the previous page we summarized these properties, which we will call "The Laws of Algebra". They are *very* important. Bookmark that page. You will be referring back to it a lot.

It is crucial to your understanding of algebra to realize that the properties stated above are what drive the symbolic manipulations for all of algebra. So it's really, really important to understand them. You may think of algebra as being made up of a lot of rules. But really, the ones we just listed are the only true "rules" that must be followed. Every other so called rule, property, or procedure you might find in algebra is just a consequence of these few that we have just listed. That's why we call them the "Laws of Algebra". If you know these laws, and follow them, you can do anything you want (or anything someone might ask you to do) in algebra. They are few but they are powerful.

We have often heard students complain that it's hard to know what you can do and what you can't do in algebra. What it boils down to is this: if it follows these Laws, you can do it. If it doesn't follow these Laws, then you can't. But wait, you may be thinking, isn't there a lot more to algebra than just this? What about things like "combining like terms?" or "cancelling", or "solving equations"? Yes, there is a lot more to algebra, but every rule, property and procedure in algebra that isn't one of these laws we've just listed is derived from these laws. In the next several sections we will be looking more closely at these and more useful properties and procedures in order to see how they follow from these laws, so you can make sense of why they work, and that way you can understand what you can and what you can't do in algebra.

Symbols for negatives and subtraction. Perhaps you've already noticed the distinction we have made between the symbol for subtraction, and the symbol to indicate negative numbers. For example, the number $^-3$ (read 'negative 3') is defined to be the number that you add to 3 to get 0. We then defined subtraction using the concept of the additive inverse again so that, for example, $5 - 3 = 5 + {}^-3$.

Typically texts do not make a typographical distinction between the dashes in the expressions $5 - 3$ and $5 + {}^-3$, even though they are mathematically quite different. In the first case, the dash is representing the *operation* of subtraction, whereas in $^-3$, the dash is part of the numeral for the number negative three. You'll note that calculators have different keys for these two concepts, since they need to make the distinction between subtraction and negative numbers. And in this text, we will also continue to make that distinction.

Homework Set 4

1) Determine the additive inverse of the following numbers. Use the definition to explain why you are correct.
   - a) 17
   - b) $^-8$
   - c) $\frac{5}{9}$
   - d) $-\frac{3}{10}$
   - e) x
   - f) $(a - b)$

2) Determine the multiplicative inverse of the following numbers. Use the definition to explain why you are correct.
    a) 17
    b) ⁻8
    c) $\frac{5}{9}$
    d) $-\frac{3}{10}$
    e) x
    f) $(a - b)$

3) Exponentiation is an operation on the set of natural numbers, since for any two natural numbers x and y, there is exactly one natural number that is x^y. Determine the properties that the operation x^y has on the set of natural numbers.
    a) Is it commutative? Make an argument based on the definition of commutativity.
    b) Is it Associative? Make an argument based on the definition of associativity.
    c) Is there an identity element? Make an argument based on the definition of an identity element.
    d) Are there inverses? Make an argument based on the definition of inverses.

4) You've probably heard this many times before: "you can't divide by zero". The more technical way of saying this is that "division by zero is undefined". In this problem we'll ask you to explain why this is a logical consequences of the definitions we've made. We defined division as multiplying by the multiplicative inverse. So dividing by zero is the same as multiplying by the multiplicative inverse of zero. Use the definition of multiplicative inverse to prove that 0 doesn't have a multiplicative inverse.

5) Fill in the table below defining an operation # on the set {w,x,y,z} so that the following conditions are met:

- **x** is the identity
- **w** and **z** are inverses
- # is commutative
- # is **not** associative

#	w	x	y	z
w				
x				
y				
z				

Be sure you can justify how each of the conditions are met by your operation.

Class Activity 5: Curses, Foiled Again!

Determine whether each of the following is true in general.

1. $(a+b)^2 = a^2 + b^2$

2. $\sqrt{a+b} = \sqrt{a} + \sqrt{b}$

3. $\frac{1}{a+b} = \frac{1}{a} + \frac{1}{b}$

4. $2^{a+b} = 2^a + 2^b$

5. $a(bc) = (ab)(ac)$

> Here is a statement of the Distributive Law:
>
> Given any real numbers a, b, and c: $a \cdot (b + c) = a \cdot b + a \cdot c$

1) Multiplying by natural numbers can be viewed as repeated addition: $3 \cdot 4 = 4 + 4 + 4$. See if you can use repeated addition to explain why the Distributive Law makes sense in the case of multiplying by a natural number. For example, why does is make sense that $5(b + c) = 5b + 5c$?

2) The Distributive Law can be extended to multiplying any number of terms. Replace a with $x + y$ in the statement above to show that

$$(x + y)(b + c) = xb + xc + yb + yc$$

3. Consider a rectangle where one side has length $x + y$ and another side has length $b + c$. Draw a careful diagram to illustrate makes $(x + y)(b + c) = xb + xc + yb + yc$ makes sense in terms of the areas of the resulting four parts of the rectangle.

Read and Study 5: The Distributive Law

"Can you do addition?" the White Queen asked. "What's one and one and one and one and one and one and one and one and one?"
"I don't know," said Alice. "I lost count."

Lewis Carrol, Through the Looking Glass

In the class activity Curses, Foiled Again!, we saw that importance of realizing that the Distributive Law applies **only** to multiplication and addition. We will say it again:

The Distributive Law applies **only** to multiplication and addition.

This is very special. Most other operations that you will come across in algebra do not "distribute". Exponents do not distribute. Square roots do not distribute. Fractions do not distribute. Furthermore, multiplication does not distribute over multiplication. We have seen students erroneously rewrite expressions like $3(xy)$ as $3x3y$. There is no such rule in math that says that whatever is outside parentheses gets distributed to everything inside the parentheses. **The *only* thing that "distributes" is multiplication over addition.**

Since this is such a huge idea, let's pause to think about why it makes sense that multiplication distributes over addition. A key concept to making sense of multiplication, especially when multiplying by natural numbers, is to think of it as "repeated addition". For example, 3 times 6 can be thought of as adding 6 together three times:

$$3 \cdot 6 = 6 + 6 + 6,$$

and, in general, 3 times a can be thought of as adding a together three times:

$$3a = a + a + a.$$

Similarly, if we have the expression $3(a + b)$, we can think of this as adding $a + b$ together three times:

$$3(a + b) = (a + b) + (a + b) + (a + b)$$

But since addition is associative and commutative, we can re-arrange the additions of the a's and b's:

$$3(a + b) = (a + b) + (a + b) + (a + b)$$

$$3(a + b) = a + a + a + b + b + b$$

$$3(a + b) = 3a + 3b$$

In other words, adding $a + b$ together three times is the same as adding three a's and three b's together.

To put this into a context, suppose a fish farm has c ponds, and each pond has exactly a salmon and b trout. How many fish are there all together? One way to answer this is to think that each pond has $a + b$ fish in it, and there are c ponds, so the number of fish is $c(a + b)$. But someone else might think this way: the number of salmon is ca, while the number of trout is cb, so the total number of fish is $ca + cb$. Hence, the distributive law is really just this idea that these two ways of counting are equivalent.

Now for a discussion of some very important terminology that will allow us to talk more precisely about algebraic expressions. In a product, the components being multiplied are called **factors**. However, in a sum, the components being added are called **terms**. That is,

> A **factor** is something that is being multiplied in a product.
> A **term** is something that is being added in a sum.

For example:
- The expression $5x$ is a product with two factors: 5 and x.
- The expression $3x^2$ is a product with two factors: 3 and x^2.
- The expression $3x^2 + 5x$ is a sum with two terms: $3x^2$ and $5x$.

Now even though 5 is a factor of the expression $5x$, please note that 5 is not a factor of the expression $3x^2 + 5x$. This is since the expression $3x^2 + 5x$ taken as a whole is not a product, it's a sum. This is a crucial distinction.

As further examples, the expression $x^2 - 2x + 4$ is a sum with three terms, while the expression $7xy$ is a product with three factors.

Consider again the following statement of the Distributive Law

$$a(b + c) = ab + ac$$

On the left side of the equation, we have a product with two factors, while on the right side we have a sum with two terms. The power of the Distributive Law is that it gives us a way to change around the orders of multiplication and addition.

The useful process we call 'factoring', that is, taking an expression such as $ab + ac$ and re-writing it in the form $a(b + c)$ is really just using the distributive law. When we use the Distributive Law to rewrite a product as a sum, this is process is usually referred to as "distributing", and when we use the Distributive Law to rewrite a sum as a product, this process is called "factoring". Factoring is the reverse, or 'undoing' of distributing, and distributing is the reverse, or 'undoing' of factoring.

The distributive law is also responsible for the useful shortcut of 'combining like terms'. For example, the expression $(x + 3)^4 - 5(x + 3)^4$ can be rewritten as $^-4(x + 3)^4$. If we write the distributive law as

$$a \cdot c + b \cdot c = (a + b) \cdot c,$$

and let a be the number 1, let b be the number $^-5$, and let c be the number $(x + 3)^4$, we will get that

$$1 \cdot (x + 3)^4 - 5 \cdot (x + 3)^4 = (1 + {}^-5) \cdot (x + 3)^4.$$

The left side of the above equation can be simplified to $(x + 3)^4 - 5(x + 3)^4$, while the right side can be simplified to $^-4(x + 3)^4$. So $(x + 3)^4 - 5(x + 3)^4 = {}^-4(x + 3)^4$.

Another way to write this proof would be to start with the expression $(x + 3)^4 - 5(x + 3)^4$ and rewrite it step by step so we can make our justifications:

$$
\begin{aligned}
(x + 3)^4 - 5(x + 3)^4 &= 1 \cdot (x + 3)^4 + {}^-5 \cdot (x + 3)^4 && \text{Multiplicative Identity;} \\
& && \text{Definition of Subtraction} \\
&= (1 + {}^-5) \cdot (x + 3)^4 && \text{Distributive Law} \\
&= {}^-4(x + 3)^4. && \text{Since } 1 + {}^-5 = {}^-4
\end{aligned}
$$

Read through our arguments again closely and make sure you understand it. In the homework, we will ask you to prove some more examples of "combining like terms" by using the distributive law.

Problem Set 5

1) For each expression below, state whether the expression as it is written is a product or a sum. If it's written as a product, state the number of factors. If it's written as a sum, state the number of terms.
    a) $3(x + 1)$
    b) $x^2 - 6x$
    c) $5x^2y$
    d) $x^3 + 5x^2 - 7x$

2) Is x a factor in the following expressions as they are written? (Yes or No).
    a) $3x$
    b) $x + 4$
    c) $x(x - 1)$
    d) $x^2 + 1$
    e) $1 + x$
    f) $(x + 1)(x + 2)$
    g) $x^3 + 5x^2 - 7x$
    h) $x(x^2 + 5x^2 - 7)$

3) In the read and study, we offered an explanation for why $a(b + c) = ab + ac$ makes using fish in ponds. Extend this analogy to explain why $(a + d)(b + c) = ab + ac + db + dc$ makes sense.

4) Use distributive law to prove that $3x + 2x$ is equivalent to $5x$.

5) Use distributive law to prove that $7x^2y - 3x^2y$ is equivalent to $4x^2y$.

6) Make appropriate substitutions into the distributive law to prove that:
   a) $a(r + s + t) = ar + as + at$
   b) $(x + y + z)(b + c) = xb + xc + yb + yc + zb + zc$

7) For each function below, determine whether $f(a + b) = f(a) + f(b)$ is valid in general (true for any a and b in the domain of f). If it is not valid, prove it by giving an example of numbers a and b so that $f(a + b) \neq f(a) + f(b)$.
   a) $f(x) = x^2$
   b) $f(x) = \sqrt{x}$
   c) $f(x) = \frac{1}{x}$
   d) $f(x) = 3x + 5$

Class Activity 6: Prove It!

In this activity, you will be asked to prove some important familiar identities for real numbers, using just the Laws of Algebra summarized in the previous Read and Study Section. The point of this activity is to see how all of the properties of algebra stem from these few laws and definitions. Your challenge is to try to make an argument (proof) that the following can be derived using only the commutative, associative, identity and inverse properties of addition and multiplication, the distributive law, and the definitions of subtraction and division.

We will start out working on proving a property together.

1. $^-1 \cdot a = {^-a}$.

We realize that this is something you have likely just taken as a fact, or just as notation, and have never before been asked to prove something like this. But simple as it is, this property isn't one of the Laws of Algebra we gave you before. But we can prove this must be true, given the Laws we were given.

A key to being able to prove a property like this is to be able to interpret what the equation is saying, namely, that **if you multiply a number by negative one, the result is the additive inverse of that number**. So to prove this, we just need to show that $^-1 \cdot a$ is the additive inverse of a. And in order to prove something is the additive inverse of a, we just need to show that it satisfies the definition of the additive inverse of a. So that's where we want you to start:

What does it mean to be the additive inverse of the number a ?

Now show that the number $^-1 \cdot a$ satisfies that definition.

So $^-1 \cdot a$ is the additive inverse of a. In other words, $^-1 \cdot a = {^-a}$, which was what we were asked to prove.

Here are two more useful properties. Your task is to try to prove these properties using only the Laws of Algebra.

2. $^-(a+b) = {^-a} + {^-b}$

Note: We know this "looks" a lot like the distributive law, but remember what we said repeatedly in the previous section: Distributive Law applies **only** to multiplication and addition. While there is an addition involved, this property does not involve multiplication. Remember the negative signs symbolize additive inverses.

3. Use the Laws of Algebra to prove the following property:

$$\frac{a}{b} + \frac{c}{b} = \frac{a+c}{b}.$$

Read and Study 6: Justifying Properties

Mathematicians are like lovers. Grant a mathematician the least principle, and he will draw from it a consequence which you must also grant him, and from this consequence another.
                                                                                    Bernard Le Bovier Fontenelle

Often students come to think of algebra as being a lot of rules to memorize. But as we talked about in a previous section, the Laws of Algebra are the only true "rules" that must be memorized. Every other so called rule, property, or procedure you might find in algebra is just a consequence of these few that we have just listed. Already in the previous section we have been able to justify the "Combining Like Terms" procedure, and "FOIL" procedure by using the Distributive Law. In this section we will continue to develop and prove useful properties and procedures regarding negative numbers and fractions. By doing so, our goal is to make sense of why these rules work and why these properties are true, so that we can understand what you can and what you can't do in algebra.

First we will list some useful properties of additive inverses. Our purpose in listing them here in one place is not for you to memorize them. Also, just because these are highlighted in a box does not mean that this is the important part of the section. In fact, the most important part of this section is the arguments we will be making to prove these properties.

Some Properties of Additive Inverses

Recall that ^-a is defined as the number such that $a + (^-a) = 0$.

For all real numbers a and b:

1. $^-1 \cdot a = {^-a}$.

2. $^-(^-a) = a$.

3. $(^-a)(^-b) = ab$.

4. $^-(a + b) = {^-a} + {^-b}$

5. If b is non-zero, $^-\left(\dfrac{a}{b}\right) = \dfrac{^-a}{b} = \dfrac{a}{^-b}$.

6. If b is non-zero, $\dfrac{^-a}{^-b} = \dfrac{a}{b}$.

Recall that we make an important distinction between the symbol for subtraction, and the symbol to indicate negative numbers. In the expression $5 - 3$, the dash is representing the *operation* of subtraction, whereas in the expression 5 + ⁻3 the dash is part of the numeral for the number negative three. So when we write 5 − 3 = 5 + ⁻3, we are **not** saying that 5 − 3 and 5 + ⁻3 mean the same thing. They mean quite different things: that 5 − 3 means we are *subtracting* two numbers (five and three), and 5 + ⁻3 means we are *adding* two numbers (five and negative three). The beauty of this equation is that these two different things are equivalent: subtracting three has the same result as adding negative three.

In the class activity, you were asked to prove property 4 on the list above: $^-(a+b) = {^-a} + {^-b}$. While we are not expecting a particular format when you write proofs, here is one that you can use as a model for how to write a clear proof.

> We are trying to prove that the additive inverse of $(a+b)$ is $(^-a + {^-b})$. To prove this, we will add the two numbers together and show the result is 0.
>
> $\quad (^-a + {^-b}) + (a+b)$
> $= {^-a} + (^-b + a) + b \qquad$ (associative law of addition)
> $= {^-a} + (a + {^-b}) + b \qquad$ (commutative law addition)
> $= (^-a + a) + (^-b + b) \qquad$ (associative law of addition)
> $= 0 + 0 \qquad\qquad\qquad\quad$ (definition of additive inverses)
> $= 0$
>
> Since $(^-a + {^-b}) + (a+b) = 0$, that means that $(^-a + {^-b})$ and $(a+b)$ are additive inverses. In other words, $^-(a+b) = {^-a} + {^-b}$.

Notice that we didn't merely give a sequence of equations. There is a good logical structure to our argument. We use words (transitions and sentences) as needed to make the structure of our argument clear to the reader. We started out by explaining what we are trying to prove and how we plan to do it. Then we justified each step and claim we made by citing a specific law or definition. At the end, we explained why the result of our calculations proves what we were trying to prove.

As we discussed in the Class Activity, we realize that this property 4 looks rather like we are "distributing" the negative signs. However, the distributive property only applies to multiplication over addition. So we can't prove this property by just by citing the distributive property. However, by using property 1 (which we also proved in the Class Activity), we can rewrite the expressions using multiplication so that the distributive property does apply. Here's what that could look like:

> $\quad ^-(a+b)$
> $= {^-1} \cdot (a+b) \qquad\qquad$ (Property 1 of Additive Inverses)
> $= {^-1} \cdot a + {^-1} \cdot b \qquad$ (Distributive Law)
> $= {^-a} + {^-b} \qquad\qquad\quad$ (Property 1 of Additive Inverses)
>
> So $^-(a+b) = {^-a} + {^-b}$

This proof is shorter and simpler. But that's because all the heavy lifting was done by Property 1. Think of our work on proving Property 1 in the class activity as a capital investment: it took some effort to do it at the time, but once it's done, it can be used whenever we want to make other things easier. Everything we figure out in mathematics opens the doors to figuring out more and more interesting and complicated things. Mathematics builds upon itself. In this chapter, we are taking the time to really understand how these foundations are developed that allow us to do really powerful things with algebra.

Let us take this opportunity to reiterate something we talked about in the last chapter, that math isn't about right answers. We've just offered two different proofs that $^-(a+b) = {}^-a + {}^-b$. Both proofs are "right", in that both make logical sense and prove the property is true. And both proofs have their own value. The first proof is nice because it uses what it means to be an additive inverse. The second proof is nice because it justifies this properties apparent connection to the Distributive Law.

So keep this in mind when you are making proofs in this and later sections: there's no right way to make a proof. There are of course wrong ways (for example, if your argument is incomplete, or illogical, or makes unsupported claims), but often many good ways to making an argument to prove something. Our goal in writing sample proofs for you in the text is not for you to copy or memorize the way we show you, but to give you experience in reading well-written logical arguments so that you will be better able to make them yourself.

Fraction Notation. In the definition of division we defined the fraction $\frac{a}{b}$ to be $a \cdot \frac{1}{b}$, where $\frac{1}{b}$ is the multiplicative inverse of b. That means the (hopefully) familiar properties of fractions can also be derived from the Laws of Algebra. We list a few here:

Some Properties of Multiplicative Inverses

Suppose a, b, c, and d are real numbers, with a, b and d non-zero as necessary.

1. $\frac{a}{a} = 1$

2. $\frac{a \cdot c}{b \cdot c} = \frac{a}{b}$

3. $\frac{a}{b} + \frac{c}{b} = \frac{a+c}{b}$

4. $\frac{a}{b} \cdot \frac{c}{d} = \frac{a \cdot c}{b \cdot d}$

5. $\frac{a}{b} \div \frac{c}{d} = \frac{a \cdot d}{b \cdot c}$

Here's an example of how one can use the definitions and Laws of Algebra to justify a given property or calculation. Let's consider property 3 of fractions listed above:

$$\frac{a}{b} + \frac{c}{b} = \frac{a+c}{b}.$$

This is a hugely important property that gives us a way to add fractions, provided the fractions have the same denominator. Basically, this justifies the "get a common denominator and add the numerators" method of adding fractions. What we will show now is how this property follows from the definitions and Laws of Algebra.

First, definition of the fraction notation is that $\frac{a}{b}$ means $a \cdot \frac{1}{b}$, where $\frac{1}{b}$ is the multiplicative inverse of b. So $\frac{a}{b} + \frac{c}{b}$ just means $a \cdot \frac{1}{b} + c \cdot \frac{1}{b}$. Now we can notice that each of these two terms $a \cdot \frac{1}{b}$ and $c \cdot \frac{1}{b}$ has a factor of $\frac{1}{b}$, so we can re-write this using the Distributive Law.

Putting this all together, we can write:

$$\begin{aligned}
\frac{a}{b} + \frac{c}{b} &= a \cdot \frac{1}{b} + c \cdot \frac{1}{b} & \text{Definition of Division} \\
&= (a+c) \cdot \frac{1}{b} & \text{Distributive Law} \\
&= \frac{a+c}{b} & \text{Definition of Division}
\end{aligned}$$

Hence, $\frac{a}{b} + \frac{c}{b} = \frac{a+c}{b}$, which was what we were trying to prove.

We will close this section with a proof of another property of fractions, that

$$\frac{1}{\left(\frac{a}{b}\right)} = \frac{b}{a}.$$

Now, you may have come to interpret the notation $\frac{1}{(\ )}$ to mean "flip" or "reciprocal", in which case this property seems obvious. But in fact there is something here to prove, since we have defined the $\frac{1}{(\ )}$ notation to mean "multiplicative inverse". So this property is saying that the multiplicative inverse of $\frac{a}{b}$ is $\frac{b}{a}$. Well, how does one prove that two numbers are multiplicative inverses? We show they satisfy the definition of multiplicative inverses, of course.

By definition, multiplicative inverses are numbers that multiply together to get 1. So we can prove that $\frac{b}{a}$ and $\frac{a}{b}$ are multiplicative inverses by multiplying them together and showing the result is 1. Here's how that argument can be written:

$$\left(\frac{b}{a}\right) \cdot \left(\frac{a}{b}\right) = \left(b \cdot \frac{1}{a}\right) \cdot \left(a \cdot \frac{1}{b}\right) \qquad \text{definition of division}$$

$$= (b) \left(\frac{1}{a} \cdot a\right) \cdot \left(\frac{1}{b}\right) \qquad \text{associative law of multiplication}$$

$$= (b) \cdot (1) \cdot \left(\frac{1}{b}\right) \qquad \text{definition of multiplicative inverses}$$

$$= b \cdot \frac{1}{b} \qquad \text{definition of multiplicative identity}$$

$$= 1 \qquad \text{definition of multiplicative inverses}$$

Since $\left(\frac{b}{a}\right) \cdot \left(\frac{a}{b}\right) = 1$, that means that $\frac{b}{a}$ and $\frac{a}{b}$ are multiplicative inverses. In other words, $\frac{1}{\left(\frac{a}{b}\right)} = \frac{b}{a}$, which was what we were trying to prove.

Homework Set 6

1) Use the definition of subtraction to explain why $a - (\bar{\ } b) = a + b$.

2) Use the Laws of Algebra to prove that $\frac{1}{(a \cdot b)} = \frac{1}{a} \cdot \frac{1}{b}$.

   Note: this equation is a statement saying that the multiplicative inverse of $(a \cdot b)$ is $\left(\frac{1}{a} \cdot \frac{1}{b}\right)$. So you are being asked to prove that the number $(a \cdot b)$ and the number $\left(\frac{1}{a} \cdot \frac{1}{b}\right)$ are multiplicative inverses.

3) Try to use the method in problem 2 above to prove that $\frac{1}{\left(\frac{1}{a} + \frac{1}{b}\right)} = a + b$. What goes wrong in your proof? Now show that $\frac{1}{\left(\frac{1}{a} + \frac{1}{b}\right)}$ and $a + b$ are not equivalent expressions by computing both expressions for several choices of the numbers a and b.

4) Use the definitions and Laws of Algebra to prove property number 4 on our list of properties of multiplicative inverses: $\frac{a}{b} \cdot \frac{c}{d} = \frac{(a \cdot c)}{(b \cdot d)}$.

5) Consider the property $\frac{1}{(\bar{\ }a)} = \bar{\ }\left(\frac{1}{a}\right)$. Note that there are two kinds of inverses involved here, additive and multiplicative.
   a) Prove that $\frac{1}{(\bar{\ }a)} = \bar{\ }\left(\frac{1}{a}\right)$ by interpreting the equation as a statement about multiplicative inverses.
   b) Prove that $\frac{1}{(\bar{\ }a)} = \bar{\ }\left(\frac{1}{a}\right)$ by interpreting the equation as a statement about additive inverses.

Class Activity 7: Cancellation Policy

1. Can you "cancel" the x's as shown in the following? If you can, prove it. If not, explain why, and give an example of values for the variables that shows the cancellation is invalid.

a. $\dfrac{5y-2x}{3x} \overset{?}{\Rightarrow} \dfrac{5y-2}{3}$

b. $\dfrac{5x-xy}{2xy} \overset{?}{\Rightarrow} \dfrac{5-y}{2y}$

c. $x + y = 3 + x + y^2 \overset{?}{\Rightarrow} y = 3 + y^2$

d. $5(y + x) = 4 + x \overset{?}{\Rightarrow} 5y = 4$

e. $5(y + x) = 4 + x \overset{?}{\Rightarrow} 5(y + 1) = 4 + 1$

2. Use the Laws of Algebra to prove that for any real numbers a, b, and c:

$$\text{If } a + c = b + c, \text{ then } a = b$$

3. Use the Laws of Algebra to prove that for any real numbers a, b, and c, (where b and c are non-zero):

$$\frac{a \cdot c}{b \cdot c} = \frac{a}{b}$$

4. Why can't you use the Laws of Algebra to prove that $\frac{a+c}{b+c} = \frac{a}{b}$? Try it!

Read and Study 7: Cancellation Facts and Myths

Everything should be made as simple as possible, but not one bit simpler.
                                                                    Albert Einstein

Much of the activity of "doing" algebra is taking an expression and rewriting it as an **equivalent expression**, often with a goal of writing them in a simpler form. Two expressions are **equivalent** if they have the same value for all relevant values of the variables involved. The big idea of this chapter is that the only things that you can do to re-write an expression as an equivalent one is to use one of the definitions or Laws of Algebra that we have presented, or one of the properties that we have proven follow from these definitions and laws.

For example, in a previous section, we used the distributive law to prove that the expression $(x+3)^4 - 5(x+3)^4$ is equivalent to the expression $^-4(x+3)^4$. Since the distributive law is true for all numbers expression $(x+3)^4 - 5(x+3)^4$ will **always** have the same value as the expression $^-4(x+3)^4$, regardless of what number x is.

However, sometimes we say two expressions are equivalent even when there might be an exception to when the expressions have the same value. For example consider the two expressions $\frac{5(x-1)^2}{x-1}$ and $5(x-1)$. We can justify these two expression are equivalent by using the definition of exponent notation as repeated multiplication, and the definition of fraction notation as multiplying by the multiplicative inverse, so that:

$$\frac{5(x-1)^2}{x-1} = \frac{5(x-1) \cdot (x-1)}{x-1} \qquad \text{Exponent notation}$$

$$= 5(x-1) \cdot (x-1) \cdot \frac{1}{(x-1)} \qquad \text{Definition of Division}$$

$$= 5(x-1) \cdot 1 \qquad \text{Definition of Multiplicative Inverses}$$

$$= 5(x-1) \qquad \text{Definition of Multiplicative Identity}$$

So it seems like we have proved that expressions $\frac{5(x-1)^2}{x-1}$ and $5(x-1)$ will have the same value for all values of x. However, there is one exception. Can you spot the one value for x for which these expressions would not have the same value?

The argument we made is valid only when the expressions involved are defined. Notice that the expression $\frac{5(x-1)^2}{x-1}$ is not defined when $x = 1$, since division by zero is not defined. However, the simplified expression $5(x-1)$ **is** defined when $x = 1$. (And it has the value of 0.) So the two expressions are not equivalent when $x = 1$. Really when we made our argument, and invoked the definition of multiplicative inverses, we should have added the provision "provided x is not 1, so that $\frac{1}{x-1}$ exists." We like to be precise, so we will do that now:

$$\begin{aligned}
\frac{5(x-1)^2}{x-1} &= \frac{5(x-1)\cdot(x-1)}{x-1} &&\text{Exponent notation}\\
&= 5(x-1)\cdot(x-1)\cdot\frac{1}{(x-1)} &&\text{Definition of Division}\\
&= 5(x-1)\cdot 1 &&\text{Definition of Multiplicative Inverses,}\\
& &&\text{provided }\frac{1}{(x-1)}\text{ exists}\\
&= 5(x-1) &&\text{Definition of Multiplicative Identity}
\end{aligned}$$

So now it's clear that our argument works for any number x except when $x = 1$. So we can say that $\frac{5(x-1)^2}{x-1}$ and $5(x-1)$ are equivalent expressions, provided $x \neq 1$.

Now we know it's common for people to describe rewriting $\frac{5(x-1)^2}{x-1}$ as $5(x-1)$ as "cancelling" the factor $(x-1)$. In our experience, "cancelling" is a trouble spot for in algebra for many students. Unfortunately, teachers, texts and students when doing algebra will often justify a step they take just by saying something like "then we cancel" and start crossing out things from both sides of an equation, or from the top and bottom of a fraction, without justifying why the cancellation is valid. The problem with this is that it's easy to get the impression that it's always OK to cancel, and that "cancellation" is some kind of mathematical law that allows you to get rid of things that are the same from the tops and bottoms of fractions, or from both sides of an equation.

Well, it's not. There is no such "cancellation" law in algebra. The truth is that "cancellation" is just a shortcut name that people have given to any process that has the ultimate effect of removing terms or factors from an expression to write an equivalent expression. Like everything in algebra, it must be able to be justified by the laws and definitions of algebra in order to be valid, just like we did in our example above.

There are two general situations that are often called "cancelling". The first is "cancelling" a common factor from the numerator and denominator of a fraction. In the Class Activity, we asked you to prove that when there is a common factor in the numerator and denominator of a fraction, that the fraction can be simplified to one without that common factor. Here's a way such an argument could be written down:

$$\frac{a\cdot c}{b\cdot c} = (a\cdot c)\cdot\frac{1}{(b\cdot c)} \qquad\text{Definition of Division}$$

$$= (a \cdot c) \cdot \left(\frac{1}{b} \cdot \frac{1}{c}\right) \qquad \text{Previous homework problem}$$

$$= \left(a \cdot \frac{1}{b}\right) \cdot \left(c \cdot \frac{1}{c}\right) \qquad \text{Associativity and Commutativity of Multiplication}$$

$$= \left(a \cdot \frac{1}{b}\right) \cdot 1 \qquad \text{Definition of Multiplicative Inverse}$$

$$= a \cdot \frac{1}{b} \qquad \text{Definition of Multiplicative Identity}$$

$$= \frac{a}{b} \qquad \text{Definition of Division}$$

Again, in math we are sticklers for precision, so we should note that the argument we made is valid only when the expressions involved are defined. In particular, b can't be zero, and neither can c.

In our argument, we tried to use only the original Laws of Algebra. But right away, it became expedient to use the fact that $\frac{1}{(b \cdot c)} = \frac{1}{b} \cdot \frac{1}{c}$ which requires a separate argument that we asked to you to make in the homework in a previous section. So even though everything in algebra can be proved ultimately from the Laws of Algebra it can become tedious to do so unless we build upon previously proven properties. That's another reason why we highlight and list certain properties that we are able to prove: so we can use them in our subsequent arguments and don't have to keep "reinventing the wheel".

With this in mind, here's another shorter proof, making use of more results from the previous section:

$$\frac{a \cdot c}{b \cdot c} = \frac{a}{b} \cdot \frac{c}{c} \qquad \text{Property 4 of Multiplicative Inverses}$$

$$= \frac{a}{b} \cdot 1 \qquad \text{Property 1 of Multiplicative Inverses}$$

$$= \frac{a}{b} \qquad \text{Definition of Multiplicative Identity}$$

So we can prove that $\frac{a \cdot c}{b \cdot c} = \frac{a}{b}$, which justifies "cancelling" the c's from the numerator and denominator. But it is crucial that we pay attention to *why* this worked. The key is that the number c was a *factor* (that is, being multiplied) of both the numerator and the denominator. Without c being a *factor*, we wouldn't be able to use Property 5 of Fractions.

Consider instead the expression $\frac{a+c}{b+c}$. Here again you might be tempted to "cancel" the number c from the numerator and denominator, resulting in the expression $\frac{a}{b}$. However, if we tried to make an argument to prove this works, we would find that there is no way to justify this using the Laws of Algebra, or any of the properties we've already proven.

Take a few minutes now to look back at all of our Laws and properties from the previous section. There is nothing there that gives us a way to simplify $\frac{a+c}{b+c}$ in a way to get rid of the c's. And this make sense because the c's are being *added*, while fractions are defined as a division, and division is defined as a *multiplication*. So ultimately we have both addition and multiplication going on here. The only Law that relates addition and multiplication is the Distributive Law. So only way to try to rewrite the expression $\frac{a+c}{b+c}$ would be to use the Distributive Law:

$$\frac{a+c}{b+c} = (a+c) \cdot \frac{1}{(b+c)} \qquad \text{Definition of Division}$$

$$= a \cdot \frac{1}{(b+c)} + c \cdot \frac{1}{(b+c)} \qquad \text{Distributive Law}$$

Now we run into a problem. If we look at the that second term $c \cdot \frac{1}{(b+c)}$, where we were hoping to get rid of the c's, by the order operations we can't multiply by c before we evaluate the grouping symbol in the fraction $\frac{1}{(b+c)}$. But we don't have a way to simplify $\frac{1}{(b+c)}$. The key reason is that $\frac{1}{(\ )}$ refers to a **multiplicative** inverse, while $(b+c)$ is **addition**. Multiplicative inverses deal with multiplication, not addition. (Notice that in a previous homework problem we *were* able to justify a simplification of $\frac{1}{(b \cdot c)}$ and that's because we were dealing with a multiplication and a multiplicative inverse.)

We know it's tempting to think perhaps that $\frac{1}{(b+c)}$ could be rewritten as $\frac{1}{b} + \frac{1}{c}$. But ask yourself why you might think that. You might be hoping that fraction bars (division) might distribute. *But the only thing that distributes is multiplication over addition. Division doesn't distribute.* Or you might think that since we were able to prove in previous homework problem that $\frac{1}{(b \cdot c)} = \frac{1}{b} \cdot \frac{1}{c}$, perhaps this will work for addition as well. Well, let's try it. Let's apply the same sort of reasoning as we did then, and see if we can prove that that $\frac{1}{(b+c)}$ is equivalent to $\frac{1}{b} + \frac{1}{c}$.

By the definition of multiplicative inverse, $\frac{1}{(b+c)}$ is the number you multiply to $(b+c)$ to get 1. To see if $\left(\frac{1}{b} + \frac{1}{c}\right)$ works, we will multiply $(b+c)$ with $\left(\frac{1}{b} + \frac{1}{c}\right)$ and see if we get 1.

$$(b+c) \cdot \left(\frac{1}{b} + \frac{1}{c}\right) = b \cdot \left(\frac{1}{b} + \frac{1}{c}\right) \ c \cdot \left(\frac{1}{b} + \frac{1}{c}\right) \qquad \text{Distributive Law}$$

$$= b \cdot \frac{1}{b} + b \cdot \frac{1}{c} + c \cdot \frac{1}{b} + c \cdot \frac{1}{c} \quad \text{Distributive Law}$$

$$= 1 + b \cdot \frac{1}{c} + c \cdot \frac{1}{b} + 1 \quad \text{Multiplicative Inverses}$$

$$= 2 + \frac{b}{c} + \frac{c}{b} \quad \text{Definition of Division}$$

Umm, this is not simplifying to 1. While we did have the b and $\frac{1}{b}$ match up get 1, we also got the c and $\frac{1}{c}$ match up get 1, and these added to 2. Plus we still have those two mixed terms remaining, and those will depend on what the numbers b and c are. For example, if b was 2 and c was 3, then $2 + \frac{b}{c} + \frac{c}{b}$ would equal $2 + \frac{2}{3} + \frac{3}{2}$, definitely not equal to 1.

Which gives us a very useful idea. When we were considering whether $\frac{1}{(b+c)}$ might be equivalent to $\left(\frac{1}{b} + \frac{1}{c}\right)$, we could have started by first checking to see if it works for some particular numbers b and c. Remember that algebraic expressions are representing **numbers.** Saying whether $\frac{1}{(b+c)}$ is equivalent to $\left(\frac{1}{b} + \frac{1}{c}\right)$ means that for whatever numbers b and c might represent, the number $\frac{1}{(b+c)}$ is the same number as $\left(\frac{1}{b} + \frac{1}{c}\right)$. And if this is not true for some particular numbers b and c, then it can't be true for all numbers b and c. If we were to try $b = 2$ and $c = 3$, then we'd find the expression $\frac{1}{(b+c)}$ would equal $\frac{1}{5}$, while the expression $\left(\frac{1}{b} + \frac{1}{c}\right)$ would equal $\frac{5}{6}$. So these expressions are not equivalent.

Now let's go back to what started this whole discussion, which was whether you can "cancel" the c's in the expression $\frac{a+c}{b+c}$. We've already discussed rather thoroughly how we are unable to justify such a cancellation by using the Laws of Algebra. But a simpler way to prove that cancelling the c's would be invalid would be just by checking it out for some particular values of a, b, and c. Suppose $a = 1$, $b = 2$, and $c = 3$, then $\frac{a+c}{b+c} = \frac{4}{5}$, while $\frac{a}{b} = \frac{1}{2}$. So clearly the c's don't cancel.

So now you might be wondering why we spent all that time in this section trying to prove something that we could have seen right away is false just by substituting in some values. We realize it's a little strange in a math book to see examples of attempted proofs that don't work, but a danger of presenting only polished finished proofs is that it can hide the reasoning that goes into trying to make an argument to prove something. Moreover, the main idea of this book is that everything in algebra makes sense and can be justified using mathematical reasoning, and it is this reasoning that tells you what you can and cannot do in algebra. So just as it's crucial to understanding why what **you can** do in algebra makes sense and can be justified by reasoning, we feel it's important to also try to understand why what **you can't** do in algebra doesn't make sense and can't be justified by reasoning.

We'll finish this section with a brief discussion of another situation that is often described as "cancelling", namely when the same expression is removed from both sides of an equation.

Suppose we have the equation $a + c = b + c$. Then we can simplify the equation as follows:

$$a + c = b + c$$

$$a + c + {}^-c = b + c + {}^-c \quad \text{Add } {}^-c \text{ to both sides of the equation}$$

$$a + 0 = b + 0 \quad \text{Definition of Additive Inverse}$$

$$a = b \quad \text{Definition of Additive Identity}$$

So if $a + c = b + c$, then $a = b$, which justifies the shortcut that people call "cancelling" the number c from both sides of the equation. We will study equations and what it means to solve equations in more detail in subsequent sections, but for now we'd just like to point out that in both situations that people describe as cancelling (simplifying fractions and solving equations), what is really going on is the use of an inverse to undo an operation. When simplifying a fraction by "cancelling" a common factor, what is really happening is we have an expression that is being multiplied by a number, then being multiplied by the inverse of that number, which undoes that first multiplication. In effect, the multiplying by the inverse "cancels" the original multiplication. Similarly, when simplifying an equation by "cancelling" a common term from each side of the equation, what is really happening is that we can add the additive inverse of that term to each side of the equation, which would undo that addition from both sides. The adding the inverse "cancels" the original addition from both sides.

Homework Set 7

In #1-4 below, when claiming two expressions are equivalent, you can ignore any values for x where you would be dividing by zero.

1) Which of these are equivalent to $\frac{x^2+1}{x}$? For each pair of expressions that are equivalent, prove you are correct with an argument based on the definitions and Laws of Algebra.
   a) $x + 1$
   b) $x + \frac{1}{x}$
   c) 1

2) Which of these are equivalent to $\frac{x}{x+1}$? For each pair of expressions that are equivalent, prove you are correct with an argument based on the definitions and Laws of Algebra.
   a) $\frac{1}{2}$
   b) 1
   c) 0

3) Which of these are equivalent to $\frac{3}{4}$? For each pair of expressions that are equivalent, prove you are correct with an argument based on the definitions and Laws of Algebra.

   a) $\frac{3-3x}{4-4x}$

   b) $\frac{x^2+3}{x^2+4}$

   c) $\frac{3+x}{4} - x$

4) Which of these are equivalent to $x - 2$? For each pair of expressions that are equivalent, prove you are correct with an argument based on the definitions and Laws of Algebra.

   a) $\frac{4x-2}{4}$

   b) $\frac{5x-10}{5}$

   c) $\frac{x^2-2}{x}$

   d) $\frac{(x-2)(x-6)}{x-6}$

   e) $\frac{(2-x)(x-6)}{6-x}$

   f) $\frac{x^2-16}{x+8}$

   g) $\frac{x^2+x-6}{x+3}$

5) Give two examples of numbers x and y that shows that $\frac{1}{x+y}$ is not equivalent to $\frac{1}{x} + \frac{1}{y}$.

6) Can you "cancel" the x's as shown? In other words, are the two expressions equivalent? (Ignore any values for x where you would be dividing by zero.) If they are, prove the two expressions are equivalent definitions and the Laws of Algebra. If not, explain why the two expressions are not equivalent and give an example of values for the variables that shows the cancellation is invalid.

   a) $\frac{5x^2-x}{x} \overset{?}{\Rightarrow} 5x^2 - 1$

   b) $\frac{5x}{2x+y} \overset{?}{\Rightarrow} \frac{5}{2+y}$

Class Activity 8: Exponentially Yours

> Let a be a positive real number and let n be a natural number. Then the **exponential** expression a^n is defined to mean a multiplied together n times. That is,
>
> $$a^n = \underbrace{a \cdot a \cdot a \cdot \ \cdots \ \cdot a}_{n \text{ times}}.$$

Use the above definition to prove the following properties of exponents hold for positive real numbers a and b, and assuming the exponents n and n are natural numbers.

1) $a^n \cdot a^m = a^{n+m}$

2) $(a^n)^m = a^{n \cdot m}$

3) $(a \cdot b)^n = a^n \cdot b^n$

Use the definition to explain why $(a + b)^n = a^n + b^n$ is NOT a property of exponents.

We define real number exponents to have the same properties that we proved are true for natural number exponents. Namely, for positive real numbers a and b, and any real number exponents x and y:

1) $a^x \cdot a^y = a^{x+y}$

2) $(a^x)^y = a^{x \cdot y}$

3) $(a \cdot b)^x = a^x \cdot b^x$

Use the properties defined above figure out the value of the following:

a. a^0 (Suggestion: use Property 1 and let $y = 0$.)

b. a^{-1} (Suggestion: use Property 1 and let $x = 1$ and $y = {-1}$.)

c. $a^{\frac{1}{2}}$ (Suggestion: use Property 1 and let $x = \frac{1}{2}$ and $y = \frac{1}{2}$.)

Read and Study 8: Exponent Definitions

The different branches of Arithmetic -- Ambition, Distraction, Uglification, and Derision.

Lewis Carroll, Alice in Wonderland

Exponents began as a short-hand notation for repeated multiplication. For example, if we want to multiply ten 2's together, we could write:

$$2 \cdot 2 \cdot 2 \cdot 2 \cdot 2 \cdot 2 \cdot 2 \cdot 2 \cdot 2 \cdot 2,$$

or we could just write:

$$2^{10},$$

where the 10 in the superscript indicates how many twos are to be multiplied together.

But if we define an exponent to be the number times the base gets multiplied together, then could something like 2^0 ever make sense? What would it mean to multiply zero twos together? Can that even be done? It sounds rather zen. If it can be done, what should the result be? 0? 1? Something else? Or what about 2^{-1}, could that ever make sense? How can you multiply twos together $^-1$ times? Could you ever have $2^{\frac{1}{2}}$? If so, how can you multiply twos together $\frac{1}{2}$ a time? Or should it be multiply only half a 2?

In mathematics, definitions are made in order for patterns to continue and be consistent. We can make definitions of exponents so that the patterns we see for positive integer exponents continue. For example, consider this pattern. We start with $7^5 = 7 \cdot 7 \cdot 7 \cdot 7 \cdot 7$, and in each line, we reduce the exponent by one:

$$
\begin{aligned}
7^5 &= 7 \cdot 7 \cdot 7 \cdot 7 \cdot 7 \\
7^4 &= 7 \cdot 7 \cdot 7 \cdot 7 \\
7^3 &= 7 \cdot 7 \cdot 7 \\
7^2 &= 7 \cdot 7 \\
7^1 &= 7
\end{aligned}
$$

Based on this pattern, what would you say 7^0 should be? Well, it depends on how you see the pattern. If you see it as "crossing off" a seven at each line, you might argue that 7^0 should be zero, as all the sevens should be crossed off. But "crossing off" is not mathematics. (And zero is not the same as "nothing".) Really, at each new line we are "undoing" one multiplication by 7, and the operation that is defined as "undoing" multiplication is division. So if we state the pattern mathematically, it's that each time the exponent decreases by one, we divide by 7, or equivalently, multiply by $\frac{1}{7}$. If we continue this pattern, we get:

$$7^5 = 7 \cdot 7 \cdot 7 \cdot 7 \cdot 7$$

$$7^4 = 7 \cdot 7 \cdot 7 \cdot 7 \qquad \text{divide by 7}$$

$$7^3 = 7 \cdot 7 \cdot 7 \qquad \text{divide by 7}$$

$$7^2 = 7 \cdot 7 \qquad \text{divide by 7}$$

$$7^1 = 7 \qquad \text{divide by 7}$$

$$7^0 = 1 \qquad \text{divide by 7}$$

$$7^{-1} = \frac{1}{7} \qquad \text{divide by 7}$$

$$7^{-2} = \frac{1}{7} \cdot \frac{1}{7} \qquad \text{divide by 7}$$

$$7^{-3} = \frac{1}{7} \cdot \frac{1}{7} \cdot \frac{1}{7} \qquad \text{divide by 7}$$

$$7^{-4} = \frac{1}{7} \cdot \frac{1}{7} \cdot \frac{1}{7} \cdot \frac{1}{7} \qquad \text{divide by 7}$$

So in order to preserve this pattern, we need $a^0 = 1$, $a^{-1} = \frac{1}{a}$, and in general $a^{-x} = \frac{1}{a^x}$.

Actually, there are *a lot* of cool patterns to exponents, and it turns out to be very useful to be able to extend these patterns to negative and even fractional exponents. To do so, we just define the properties of exponents that we proved work for natural number exponents to work for real number exponents, and see what the consequences are.

In the next box we list several properties of exponents. The first three we proved with natural number exponents in the class activity. The rest we will ask you to prove in the homework. Our purpose in listing these properties here in one place is not for you to memorize them. Also, just because these are highlighted in a box does not mean that this is the important part of the section. In fact, the most important part of this section are the arguments that prove these properties. The big idea is that you can make sense of why each of these properties is true based on the definition of natural number exponents as being repeated multiplication, and then make sense of negative and fractional exponents in a way that is consistent with these properties.

Some Properties of Exponential Expressions

For all real numbers $a > 0$, $b > 0$, and all real x and y:

1) $a^x \cdot a^y = a^{x+y}$

2) $(a^x)^y = a^{x \cdot y}$

3) $(a \cdot b)^x = a^x \cdot b^x$

4) $a^{-x} = \dfrac{1}{a^x}$

5) $\dfrac{a^x}{a^y} = a^{x-y}$

6) $\left(\dfrac{a}{b}\right)^x = \dfrac{a^x}{b^x}$

We will take this opportunity to make a warning about "distributing". Consider property 3 listed above: $(a \cdot b)^x = a^x \cdot b^x$. You might be tempted to incorrectly refer to this property as 'distributing' the exponent x to each factor, but this is very dangerous thinking. This is **not** a distributive property. (Remember that the only thing that distributes is multiplication over addition!). The reason why it's dangerous to think of this property as 'distributing exponents' is that in fact, **exponents do not distribute**. For example, $(a + b)^2$ is **not** equivalent to $a^2 + b^2$. (Try it for some various values of a and b and see). The same goes for other powers.

We will close with a short discussion about terminology. In general, the terms "exponent" and "power" refer to the same thing. In the expression 7^3 we say that 7 is the base for the exponent 3. We can also call this expression 7 to the 3rd power.

However, when describing functions, the choice of the term "exponential function" or "power function" depends on whether it is the base or the exponent that is the variable. For example, $f(x) = x^3$ would be called a **power function**, whereas $f(x) = 7^x$ would be called an **exponential function.** These two classes of functions have differing characteristics and behavior and we will study power functions and exponential functions in more detail in later. But when we do, it will be the definition and properties of exponents we have explored in this section that will determine the characteristics and behavior of these two different classes of functions that involve exponents.

Homework Set 8

In 1-4, determine which expressions are equivalent (you can ignore any values where you would be dividing by zero.)

1) Which of these are equivalent to $8x^3$? For each pair that is equivalent, make an argument using the definitions and Laws of Algebra.

   a) $2^3 \cdot x \cdot x^2$

   b) $(2x)^3$

   c) $\frac{16x^{12}}{2x^4}$

   d) $\frac{8}{x^{-3}}$

2) Which of these are equivalent to $\frac{y^2}{y^6}$? For each pair that is equivalent, make an argument using the definitions and Laws of Algebra.

   a) $y^{\frac{1}{3}}$

   b) y^{-4}

   c) $\frac{1}{3}$

   d) $y^2 \cdot y^{-6}$

3) Which of the following are equivalent to 6^x ? For each pair that is equivalent, make an argument using the definitions and Laws of Algebra.

   a) $2^x \cdot 3^x$

   b) $\frac{12^x}{2^x}$

   c) $2^x + 4^x$

   d) $(^-6)^{-x}$

4) Which of the following are equivalent to $2 \cdot 9^x$? For each pair that is equivalent, make an argument using the definitions and Laws of Algebra.

   a) $(2^{-x} \cdot 9)^x$

   b) $2 \cdot (3^x)^2$

   c) $\left(\frac{9^{-x}}{2}\right)^{-1}$

   d) 18^x

5) Find the value(s) for the variable x that makes these equations true. **Do not use a calculator**. Use the definition and properties of exponents to figure these out.

   a) $3^x = 81$

   b) $3^x = \frac{1}{27}$

   c) $x^3 = 64$

   d) $x^2 = 64$

   e) $2^x = 64$

   f) $2^x = \frac{1}{8}$

6) Which of the following are equivalent to 3?

   a) $\left(\frac{1}{3}\right)^{-1}$

   b) $-\left(\frac{1}{3}\right)$

   c) $-(^-3)$

   d) $(^-3)^{-1}$

   e) $\left(-\frac{1}{3}\right)^{-1}$

   f) $\frac{1}{3^{-1}}$

   g) $\frac{3^{-1}}{^-1}$

   h) $-\left(\frac{1}{-3}\right)^{-1}$

7) Use the repeated multiplication definition to justify the following properties of exponents when x and y are positive integers. Make the argument with an example with specific values for x and y (such as $x = 5$ and $y = 3$) then generalize the argument to be valid for any positive integer values for x and y.

   a. $a^x \cdot a^y = a^{x+y}$

   b. $(a^x)^y = a^{xy}$

   c. $(ab)^x = a^x b^x$

   d. $\left(\frac{a}{b}\right)^x = \frac{a^x}{b^x}$

8) Is the following a general property of exponents? If so, make an argument using the definitions and Laws of Algebra. If not, find two examples of values for a, b and x that make the equation false.
$$(a + b)^x = a^x + b^x$$

9) Prove the following properties follow from $a^x \cdot a^y = a^{x+y}$.

   a. $a^0 = 1$ Suggestion: substitute $y = 0$ into $a^x \cdot a^y = a^{x+y}$

   b. $a^{-x} = \frac{1}{a^x}$ Suggestion: substitute $y = {}^-x$ into $a^x \cdot a^y = a^{x+y}$

   c. $a^{x-y} = \frac{a^x}{a^y}$ Suggestion: substitute $y = {}^-y$ into $a^x \cdot a^y = a^{x+y}$ and use what you proved in part b.

Class Activity 9: Let's Get Radical

> The **square root of x** is defined as the non-negative number you square to get x.
> That is,
> $$\sqrt{x} = y \text{ means that } y \text{ is the non-negative number such that } (y)^2 = x.$$
>
> In general, the **n^{th} root of x** is the number you raise to the nth power to get x.
> That is,
> $$\sqrt[n]{x} = y \text{ means that } y \text{ is the number such that } (y)^n = x.$$
>
> If there is only one such number, it can be positive or negative. But if there are two such numbers, then the nth root is taken to be the non-negative one.

1. Discuss how to use definitions above (*not a calculator*) to figure out the following. Use the definitions to argue you are correct.
   a. $\sqrt{196}$
   b. $\sqrt{(-5)^2}$
   c. $\sqrt[5]{-32}$
   d. $\sqrt[3]{64}$
   e. $\sqrt[3]{(-7)^3}$

2. If possible, simplify the following expressions. Prove you are correct using the definition of the root.
   a. $\left(\sqrt{a}\right)^2$
   b. $\sqrt{(x-1)^2}$
   c. $\sqrt{t^2 - 10t + 25}$
   d. $\sqrt[3]{27y^{12}}$

3. Use the definition of the square root to prove that $\sqrt{ab} = \sqrt{a}\sqrt{b}$

4. Prove that $x^{\frac{1}{3}} = \sqrt[3]{x}$.

Read and Study 9: Roots of Numbers

If people do not believe that mathematics is simple, it is only because they do not realize how complicated life is.

John Louis von Neumann

Just as subtraction and division were defined in terms of the inverse of addition and multiplication, respectively, we define roots by undoing natural number exponents (powers).

> We define the nth **root of** x to be the number you raise to the nth power to get x. That is,
> $$\sqrt[n]{x} = y \text{ means that } y \text{ is the number such that } (y)^n = x.$$
> If there are two such numbers, then the nth root is taken to be the non-negative one.

Sometimes, you might see this definition written out in cases, depending on whether x is positive or negative. The reason for these extra technicalities related to positive and cases is that since a function is such a powerful concept, we really want the nth **root of** x to be a function. And a key part of the definition of a function as that there is one and only one output. So when there are two possible numbers, we have just decided to choose the positive number to be the output of the function.

Here's an example to illustrate. Consider the 2nd root of 9, which is more commonly referred to as the square root of 9. By definition,
$$\sqrt{9} = y \text{ means that } y \text{ is the number such that } (y)^2 = 9.$$

In other words, the square root of 9 is defined as the number you square to get 9. But there are two numbers you can square to get 9, since
$$(3)^2 = 9, \text{ and}$$
$$(^-3)^2 = 9.$$

Thinking about function machines, there are two different inputs into the squaring machine that give you an output of 9.

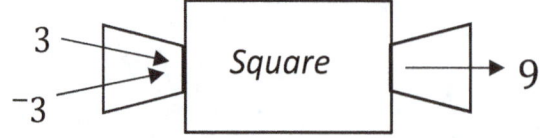

Recall that in the definition of a function, it's OK to have two different inputs giving you the same output. Now if we want to define an inverse function that takes an input of 9 and gives an output that is the square root of 9, we have a problem: functions cannot have two different outputs for the same input.

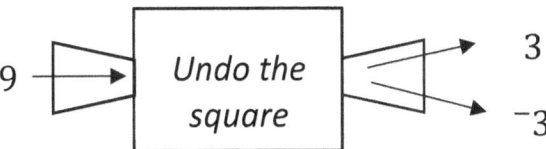

Since there's two different output for the same input, the process in the diagram above cannot be called a function. To make the "Undoing the square" process into a function, we need to pick only one of these outputs. We (as a mathematical community) have just decided to all agree to pick the positive output. (Maybe because we are such positive thinkers. Ok, that's pretty lame.)

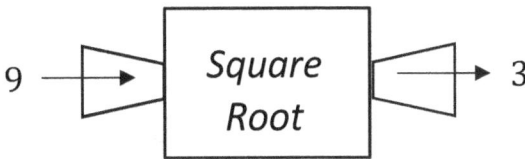

So we have only one number that we say is "the square root of 9", and that number is 3. This discussion should not be leave you with the impression that n^{th} roots are always positive. In cases where there is only one possible output to undoing the power, then the nth root can be negative. For example, let's consider the 3rd, or cube root of $^-64$, which can be written as $\sqrt[3]{^-64}$. Then according to the definition of the nth root of x,

$$\sqrt[3]{^-64} = y \text{ means that } y \text{ is the number such that } (y)^3 = {^-64}.$$

That is, $\sqrt[3]{^-64}$ is the number you cube to get $^-64$. Well there is only one such number, namely $^-4$. Check it out yourself: $(^-4)^3 = {^-4} \cdot {^-4} \cdot {^-4} = {^-64}$. (Notice that in this case positive 4 doesn't work, since $(4)^3 = 4 \cdot 4 \cdot 4 = 64$, not the required $^-64$.) Since this negative number is the only option, we take it, and write that $\sqrt[3]{^-64} = {^-4}$. So you see an nth root can be negative. As a homework problem, we will ask you to think about classifying the various cases for n and x when the nth root of x will be either one positive number or one negative number.

In the previous section, we introduced a definition for natural number exponents and used it to develop some properties for exponents, which we can then extend to other real number exponents. In the last section, we discussed how to make sense of negative exponents, and now we will see if we can extend those properties to fractional exponents.

In particular, consider the first property that we proved works for natural number exponents, namely that $a^{x+y} = a^x \cdot a^y$. Let's see what this property would require of $9^{\frac{1}{2}}$. Well, if we made a table showing the pattern with integer exponents, we might argue that $9^{\frac{1}{2}}$ should be a number between $9^0 = 1$ and $9^1 = 9$. Well, ½ is halfway between 0 and 1, so maybe $9^{\frac{1}{2}}$ should be halfway between 1 and 9. That would be 5. Let's see if $9^{\frac{1}{2}}$ being 5 would be consistent with the definition. Suppose we let $a = 9$, $x = \frac{1}{2}$ and $y = \frac{1}{2}$ in the property $a^{x+y} = a^x \cdot a^y$. Then

$$9^{\frac{1}{2}+\frac{1}{2}} = 9^{\frac{1}{2}} \cdot 9^{\frac{1}{2}} \qquad \text{Exponent Property 1}$$

$$9^1 = 9^{\frac{1}{2}} \cdot 9^{\frac{1}{2}} \qquad \text{Since } \frac{1}{2}+\frac{1}{2}=1$$

$$9 = 9^{\frac{1}{2}} \cdot 9^{\frac{1}{2}} \qquad \text{Since } 9^1 = 9$$

So if $9^{\frac{1}{2}}$ were 5, then we'd have $9 = 5 \cdot 5$, but that's not right. We need $9^{\frac{1}{2}}$ to be something that when you multiply it to itself, you get 9. So $9^{\frac{1}{2}}$ must be 3.

If we were to make this same argument for a general base a, the Exponent Property 1 implies that $a^{\frac{1}{2}} \cdot a^{\frac{1}{2}} = a$. This means that $a^{\frac{1}{2}}$ is the number you square to get a. In other words, $a^{\frac{1}{2}}$ must be $\sqrt{a}$. In general, we have the following:

Relationship between Roots and Exponents (Powers)

Whenever the following quantiles are defined,

$$a^{\frac{1}{n}} = \sqrt[n]{a}.$$

We have discovered something really really cool here. We've made a definition for exponents, and then a definition for roots, and now we see a fractional exponent is the same thing as a root! Actually, it should not be very surprising that there is a connection between roots and exponents, since roots were defined in terms of exponents ("undoing" a power). But this connection between exponents and roots is very exciting, since we have already been able to come up with lots of properties that describe how exponents behave, we can now use those properties to describe how roots behave.

Some Properties of Roots

Let a and b be real, and m and n be natural numbers. If $\sqrt[n]{a}$ and $\sqrt[n]{b}$ are real, then

1. $\sqrt[n]{a \cdot b} = \sqrt[n]{a} \cdot \sqrt[n]{b}$
2. $\sqrt[n]{\frac{a}{b}} = \frac{\sqrt[n]{a}}{\sqrt[n]{b}}$
3. $\sqrt[n]{a^m} = \left(\sqrt[n]{a}\right)^m$

Warning: n^{th} roots do not distribute!!! We know that students often try to memorize the properties above by thinking of them in terms of distributing the roots. But this is wrong and dangerous. In general $\sqrt[n]{a+b} \neq \sqrt[n]{a} + \sqrt[n]{b}$. (Try it with $\sqrt{9+16}$ and see what happens!) Remember that the distributive property is ONLY for multiplication over addition. So please, please don't try to think of the properties above in terms of distributing.

Remember that everything in algebra makes sense, and can be justified by the definitions and logical reasoning. So let's do that now to explain why these properties are true and make sense. We will leave property 1 and 2 above for you to justify in the homework, and do property 3 together, since the argument is a little more complicated.

We want to prove that $\sqrt[n]{a^m} = \left(\sqrt[n]{a}\right)^m$. Let's do this by starting with $\sqrt[n]{a^m}$, and use the definitions to show how this can be re-written as $\left(\sqrt[n]{a}\right)^m$.

$$
\begin{aligned}
\sqrt[n]{a^m} &= (a^m)^{\frac{1}{n}} && \text{Relationship between roots and exponents} \\
&= a^{m \cdot \frac{1}{n}} && \text{Property of Exponents \#2} \\
&= a^{\frac{1}{n} \cdot m} && \text{Commutativity of Multiplication} \\
&= \left(a^{\frac{1}{n}}\right)^m && \text{Property of Exponents \#2} \\
&= \left(\sqrt[n]{a}\right)^m && \text{Relationship between roots and exponents}
\end{aligned}
$$

In previous sections we have shown the power of patterned lists. We made lists of multiplications by successively lower number so that students can see patterns and how it would make sense to define multiplying by negative numbers. We also made lists of powers of a number with successively lower exponents to see patterns and how it would make sense to define negative exponents. We can further expand upon those lists to make sense of fractional exponents and roots. For example, let's consider the number 8. We could start with a list of successively higher integer powers of 8 such as:

$$
\begin{array}{rcll}
8^0 &=& 1 & \qquad 1 \\
8^1 &=& 8 & \qquad 8 \\
8^2 &=& 8 \cdot 8 & \qquad 64 \\
8^3 &=& 8 \cdot 8 \cdot 8 & \qquad 512 \\
8^4 &=& 8 \cdot 8 \cdot 8 \cdot 8 & \qquad 4{,}096
\end{array}
$$

Now if we were to try to make a guess as to what the number $8^{\frac{7}{3}}$ should be, we might argue that since 7/3 is bigger than 2 but less than 3, that $8^{\frac{4}{3}}$ should be between 64 and 512. But where in between? Maybe it should be 1/3 of the way between 64 and 512? But would that be consistent with the definitions for exponent? Realizing that fractional powers should be in between the integer powers in our table, let's insert the following lines to our table:

$$8^0 = 1 \qquad\qquad 1$$
$$8^{\frac{1}{3}}$$
$$8^{\frac{2}{3}}$$
$$8^1 = 8 \qquad\qquad 8$$
$$8^{\frac{4}{3}}$$
$$8^{\frac{5}{3}}$$
$$8^2 = 8 \cdot 8 \qquad\qquad 64$$
$$8^{\frac{7}{3}}$$
$$8^{\frac{8}{3}}$$
$$8^3 = 8 \cdot 8 \cdot 8 \qquad\qquad 512$$
$$8^{\frac{10}{3}}$$
$$8^{\frac{11}{3}}$$
$$8^4 = 8 \cdot 8 \cdot 8 \cdot 8 \qquad\qquad 4{,}096$$

Now a really neat thing about 8 is that it is 2^3. So $8 = 2 \cdot 2 \cdot 2$. Let's see what happens when we replace those 8's in the table with $2 \cdot 2 \cdot 2$ instead. We'd get

$$8^0 = 1 \qquad\qquad 1$$
$$8^{\frac{1}{3}}$$
$$8^{\frac{2}{3}}$$
$$8^1 = 2 \cdot 2 \cdot 2 \qquad\qquad 8$$
$$8^{\frac{4}{3}}$$
$$8^{\frac{5}{3}}$$
$$8^2 = 2 \cdot 2 \cdot 2 \cdot 2 \cdot 2 \cdot 2 \qquad\qquad 64$$
$$8^{\frac{7}{3}}$$
$$8^{\frac{8}{3}}$$
$$8^3 = 2 \cdot 2 \cdot 2 \cdot 2 \cdot 2 \cdot 2 \cdot 2 \cdot 2 \cdot 2 \qquad\qquad 512$$
$$8^{\frac{10}{3}}$$
$$8^{\frac{11}{3}}$$
$$8^4 = 2 \cdot 2 \cdot 2 \cdot 2 \cdot 2 \cdot 2 \cdot 2 \cdot 2 \cdot 2 \cdot 2 \cdot 2 \cdot 2 \qquad\qquad 4{,}096$$

Now what would make sense to fill in for the missing powers of 8? Go ahead and fill in the table with these fractional powers.

Homework Set 9

1) Evaluate the following. Do not use any root or exponent buttons on a calculator. Instead, use the definitions and properties of roots and exponents to figure these out.

   a) $\sqrt{(-17)^2}$

   b) $\sqrt[3]{(-17)^3}$

   c) $\sqrt[4]{(-17)^4}$

   d) $\sqrt[5]{(-17)^5}$

   e) $\sqrt[5]{-32}$

   f) $64^{\frac{1}{3}}$

   g) $64^{\frac{1}{6}}$

   h) $(-64)^{\frac{1}{3}}$

   i) $(-64)^{\frac{1}{6}}$

2) Evaluate the following. Do not use any root or exponent buttons on a calculator. Instead, use the definitions and properties of roots and exponents to figure these out.

   a) $64^{\frac{3}{2}}$

   b) $64^{\frac{2}{3}}$

   c) $(-64)^{\frac{2}{3}}$

   d) $81^{\frac{3}{4}}$

   e) $(-125)^{\frac{4}{3}}$

3) Find the value(s) for the variable x that makes these equations true. **Do not use a calculator.** Use the definition and properties of exponents and roots to figure these out.

   a. $x^3 = 8$

   b. $x^{\frac{1}{3}} = 2$

   c. $x^{\frac{1}{2}} = 8$

   d. $x^{\frac{1}{3}} = 4$

   e. $\sqrt{x} = 4$

   f. $x^{\frac{2}{3}} = 4$

   g. $x^{\frac{3}{2}} = 27$

   h. $x^{-\frac{1}{2}} = \frac{1}{3}$

   i. $\sqrt[3]{x^2} = 16$

   j. $\sqrt{x^3} = 125$

In 4-6, determine which expressions are equivalent, whenever both expressions are defined. For each pair that is equivalent, make an argument using the definitions and Laws of Algebra.

4) Which of these are equivalent to $\sqrt{4x^3}$? For each pair that is equivalent, make an argument using the definitions and Laws of Algebra.
   a) $4 \cdot \sqrt{x^3}$
   b) $2x^3$
   c) $\sqrt{4x^3 + 1} - 1$
   d) $2(\sqrt{x})^3$

5) Which of these are equivalent to $\sqrt{x^2 + 6x + 9}$? For each pair that is equivalent, make an argument using the definitions and Laws of Algebra.
   a) $x + \sqrt{6x} + 3$
   b) $x + 3$
   c) $7x + 9$
   d) $7x + 3$

6) Which of these are equivalent to $(\sqrt{x} + \sqrt{3})^2$? For each pair that is equivalent, make an argument using the definitions and Laws of Algebra.
   a) $x + 3$
   b) $(\sqrt{x})^2 + (\sqrt{3})^2$
   c) $x + 2\sqrt{3x} + 3$
   d) $(1 + 2\sqrt{3})x + 3$

7) Use the definition of the nth root to prove the following properties.
   a) $\sqrt[3]{a \cdot b} = \sqrt[3]{a} \cdot \sqrt[3]{b}$
   b) $\sqrt[3]{\dfrac{a}{b}} = \dfrac{\sqrt[3]{a}}{\sqrt[3]{b}}$

8) Determine whether each of the following are True for all real values of x and y; True for some values of x and y, and false for others; or False for all real values of x and y.

   a) $\sqrt{x^2} = x$
   b) $(\sqrt{x})^2 = x$
   c) $\dfrac{\sqrt{x}}{\sqrt{y}} = \dfrac{x}{y}$
   d) $\sqrt{x + y} = \sqrt{x} + \sqrt{y}$
   e) $\sqrt[3]{x + y} = \sqrt[3]{x} + \sqrt[3]{y}$

Class Activity 10a: Now You're Just Being Irrational!

1. Find a decimal approximation for $\sqrt{2}$, by using a calculator that does **not** have a square root button. Your approximation should be accurate to three decimal places.

2. Now use a calculator's square root button to find an approximation for $\sqrt{2}$.

It's important to realize that the number you wrote down in part 2 is also just an approximation. You can demonstrate this by using a high precision calculator (such as on a computer spreadsheet) to square the number you got in part 2 and see that the result isn't quite 2.

Rational numbers are numbers that can be written as a fraction (ratio) of integers, which also means they can be written as a repeating or terminating decimal. If a real number can not be written as a fraction of integers, it is called **irrational.** The Pythagoreans (followers of the ancient Greek philosopher Pythagoras) proved that $\sqrt{2}$ is irrational about 2500 years ago. (According to legend, a guy named Hippasus proved it, and was subsequently drowned at sea by the gods for revealing such as shocking discovery.)

But even though there is no way to write down $\sqrt{2}$ exactly as a decimal, that doesn't mean it's not an exact real number. The number $\sqrt{2}$ is very real. Let's demonstrate that now:

3. Use the straight edge of another piece of paper to draw a line segment on your paper representing a line of length 1. Now figure out how you can you construct a line segment that has length $\sqrt{2}$? (Note: you do not need a ruler!)

Class Activity 10b: Me, Myself and *i*

Why are our days numbered and not, say, lettered?

Woody Allen

Define the number i to the number you square to get $^-1$. That is, $i = \sqrt{^-1}$.

Define the set of **complex numbers** to be all numbers of the form $a + bi$, where a and b are real numbers.

1. First, use the definition to explain why each of the following are complex numbers:
   a) 7
   b) $3i$
   c) $5 - i$
   d) $\sqrt{2}$
   e) $-\frac{1}{2} + \sqrt{3}\, i$

2. Now your task now is to show that by using the definitions and Laws of Algebra, each of the following expressions can be simplified to a complex number in the form $a + bi$.

   a) $\sqrt{^-16}$
   b) $\sqrt{1-9}$
   c) $(3 - 2i) \cdot i$
   d) $(5i)^2$
   e) $(3 + 5i) \cdot (4 + 2i)$
   f) i^{-1}
   g) $\frac{1}{2+i}$
   h) $\sqrt{i}$

Read and Study 10: Irrational and Imaginary Numbers

Numbers constitute the only universal language.

Nathaniel West

In the first Class Activity for this section, you explored the number $\sqrt{2}$. By definition, this is the number you square to get 2. At first, you might be inclined to think that there is no such number. And in fact, there is no way to represent this number exactly with a fraction or decimal, so it's not a "rational number". But $\sqrt{2}$ is indeed a number. It's a "real number", in that it represents an exact distance, and as such it is point on the number line. Moreover, we can do calculations consistent with the Laws of Algebra with numbers such as $\sqrt{2}$ just as we can with numbers like 3 or $\frac{1}{2}$ or $^-7.4$.

In the second Class Activity for this section, you explored the number $\sqrt{-1}$. By definition, this is the number you square to get $^-1$. At first, you might be inclined to think that there is no such number. And in fact, there is no way to represent this number on the number line, so it's not a "real number". But $\sqrt{-1}$ is indeed a number. It's a "complex number". The reason why we call it a number is that it behaves just like a number, and we can do calculations consistent with the Laws of Algebra with numbers such as $\sqrt{-1}$ just as we can with numbers like 3 or $\frac{1}{2}$ or $^-7.4$ or $\sqrt{2}$. (Compare this last paragraph to the one preceding it. See the parallels?)

In the beginning of this book, we said that algebra is really the study of operations defined on sets. In algebra, the sets are sets of numbers, usually the real numbers, unless we have some desire to restrict our numbers to a smaller set (such as integers) or a larger set (such as complex numbers). The operations we study are basically addition and multiplication, or defined in terms of addition or multiplication, such as subtraction, division, exponents, and roots. In this read and study, we will lead you through a quasi-historical development of different sets of numbers and these operations on those sets, so you can see how these sets of numbers and these operations developed hand in hand. By doing this, we can better understand how these operations are related, and better understand the role of these different number sets.

In the beginning..., we start with the "counting", or "natural" numbers:

$$0, 1, 2, 3, 4, 5, 6, \ldots.$$

and an operation "addition" for combining them, for example

$$2 + 3 = 5.$$

Then we can think about "solving equations", that is, figuring out a missing number. For example, we can ask what number do you add to 4 to get 10?

$$x + 4 = 10.$$

We end up wanting to solve equations like this a lot, so we define a new operation, "subtraction", that gives you the answer to that question. (Saying $10 - 4 = 6$ is just another way to say that 6 is the number you add to 4 to get 10.) But then we can write equations that don't have a solution! For example, what number do you add to 4 to get 1?

$$x + 4 = 1.$$

There is no number (yet) that works. So we invent *new* numbers. For example, we'll say that $^-3$ is the number you add to 4 to get 1. In order for all equations using addition to have solutions, we've had to expand our set of numbers to the set of "integers":

$$\ldots, ^-4, ^-3, ^-2, ^-1, 0, 1, 2, 3, 4, \ldots$$

And everything is fine.

But then we come up with a new operation, multiplication, which comes from repeated addition, for example:

$$4 \cdot 3 = 3 + 3 + 3 + 3 = 12.$$

Again, we can think about solving equations using multiplication. For example, 5 times what number gives you 30?

$$5 \cdot x = 30.$$

We end up wanting to solve equations like this a lot, so we define a new operation, "division", that gives you the answer to that question. (Saying $30 \div 5 = 6$ is just another way to say that 6 is the number you multiply by 5 to get 30.) But then we can write equations using multiplication that don't have a solution! For example, what number do you multiply by to 5 to get 2?

$$5 \cdot x = 2.$$

There is no number (yet) that works. So we invent *new* numbers. For example, we'll say that $\frac{2}{5}$ is the number you multiply by 5 to get 2. Now all equations using multiplication will have solutions. (Well, almost all. The only exceptions are equations such as $0 \cdot x = 5$, which is why division by zero is "undefined"). But we've had to expand our set of numbers to the set of "rational numbers", which includes all of the integers and all fractions of integers.

And everything is fine.

But then we come up with a new operation, exponentiation, or raising to a power, which comes from repeated multiplication, for example:

$$3^4 = 3 \cdot 3 \cdot 3 \cdot 3 = 81.$$

Now we can think about equations using powers. For example, what number squared gives you 16?
$$x^2 = 16.$$

We end up wanting to solve equations like this a lot, so we define a new operation, "taking a root", that gives you the answer to that question. (Saying $\sqrt{16} = 4$ is just another way to say that 4 is the positive number you square to get 16.) But then we can write equations using powers that don't have a solution! For example, what number do you square to get 3?
$$x^2 = 3.$$

There is no number (yet) that works. So we invent *new* numbers. For example, we'll say $\sqrt{3}$ is the number you square to get 3. We've had to expand our set of numbers to the set of "real numbers", which can be thought of the set of all points on the number line, or the set of all positive or negative distances.

But there are *still* some equations involving powers that don't have solutions in the real numbers, for example the equation $x^2 = {}^-1$. So we invent *new* numbers. For example, we'll say $i = \sqrt{{}^-1}$ is the number you square to get $^-1$. Now all equations involving powers will have solution, but we've had to expand our set of numbers to the set of "complex numbers".

And everything is fine.

In the box below, we will summarize the number sets we have developed in this story. Notice how each new set of numbers in larger than the previous, in that it contains all the numbers in the previous set and introduces some important new kind of number.

Definitions of Number Sets

- **Natural Numbers**: The set of "counting numbers":
  $$1, 2, 3, 4, \ldots$$

- **Integers**: The set of natural numbers, together with all of their additive inverses and zero.
  $$\ldots, {}^-4, {}^-3, {}^-2, {}^-1, 0, 1, 2, 3, 4, \ldots$$

- **Rational Numbers**: The set of all integers, together with all of their multiplicative inverses.

- **Real Numbers**: The set of all distances and their additive inverses. This includes all rational numbers, together with all irrational numbers.

- **Complex Numbers**: The set of all real numbers, together with $i = \sqrt{{}^-1}$, and any sum or product of these numbers.

Unless we have explicitly stated otherwise, all of the definitions, laws and properties of algebra we have discussed already in this text apply to all of these kinds of numbers. If the definitions and laws of algebra work for it, then it's a number. That's why all these numbers, even irrational and complex numbers, are considered "numbers".

But it wasn't always the case. Numbers like $i = \sqrt{-1}$ were not always considered to be really numbers, hence the term "imaginary numbers" (numbers that involve i) and "real numbers" (numbers that don't involve i). In fact, the letter i caught on to represent $\sqrt{-1}$ because imaginary starts with the letter i.

But really all numbers are "imaginary" in that we as humans have thought them up in our minds. The number 7 is the result of our imagination. You may protest, saying the number 7 is real since I can have 7 coconuts. But if you look at a pile with seven coconuts, it's really just a pile of coconuts. WE are the ones that came up with this concept of "seven" as an attribute of that pile, that we can then apply the same attribute of "seven-ness" to a pile of pennies, or length of time such (such as 7 days), or distance (such as 7 miles).

In the early days of algebra (in the 1200-1500's) negative numbers used to be considered "fake" as well. Mathematicians tried hard to avoid the necessity of negative numbers, and we only came around after a few bold people showed that negative numbers behave just like positive numbers, and how using negative numbers can make solving problems much easier. To reconcile negatives as being numbers, we had to expand our concept of what a number can represent: numbers represent not only quantities of objects, but they can represent missing quantities of objects.

Similarly, with the advent of rational numbers, numbers can represent not only whole objects, but fractional parts of objects. The ancient Greeks (who gave us such mathematical titans as Euclid and Pythagoras) did not consider fractions to be numbers. But eventually we discovered that we can add, subtract, multiply and divide quantities like "halves" and "sevenths" just as well as we can with natural numbers. Then we discovered that there are quantities (such as the length of the diagonal of a square) that cannot be written as fractions (and so are irrational) but behave like numbers as well.

The modern mathematical view is summed up like this: "if it acts like a number, then it's a number". Negatives, fractions, irrationals, and imaginary numbers are all considered numbers because they all can be added and multiplied following the same Laws of Algebra.

In a previous section, we explored how negatives and fractions can be added and multiplied following the Laws of Algebra, and then we explored how irrational numbers such as $\sqrt{2}$ can be added and multiplied following these same Laws of Algebra. Now we will explore how imaginary, (and in general, complex) numbers can be added and multiplied following the same Laws of Algebra. Remember, i is a numeral representing a specific number, just like 7, $\frac{2}{3}$, or $\sqrt{2}$ are all numerals representing specific numbers.

For example, consider the complex numbers $2i$ and $5i$. Then using the Laws of Algebra, we can add these two numbers together as follows:

$$2i + 5i$$

$$= (2+5)i \qquad \text{Distributive Law}$$

$$= 7i$$

So the result is we can add $2i$ and $5i$ and get the result $7i$, which is another complex number.

And we can multiply $2i$ and $5i$ together as follows:

$$(2i) \cdot (5i)$$

$$= 2 \cdot (i \cdot 5) \cdot i \qquad \text{Associative Law of Multiplication}$$

$$= 2 \cdot (5 \cdot i) \cdot i \qquad \text{Commutative Law of Multiplication}$$

$$= (2 \cdot 5) \cdot (i \cdot i) \qquad \text{Associative Law of Multiplication}$$

$$= (10) \cdot (^{-}1) \qquad \text{Definition of the number } i$$

$$= {}^{-}10$$

As another example, consider the complex numbers $4 + 3i$ and $5 - 2i$. Then using the Laws of Algebra, we can add them together to get

$$(4 + 3i) + (5 - 2i)$$

$$= 4 + (3i + 5) - 2i \qquad \text{Associative Law of Addition}$$

$$= 4 + (5 + 3i) - 2i \qquad \text{Commutative Law of Addition}$$

$$= (4 + 5) + (3i - 2i) \qquad \text{Associative Law of Addition}$$

$$= 9 + (3 - 2)i \qquad 4 + 5 = 9; \text{ Distributive Law}$$

$$= 9 + i \qquad 3 - 2 = 1$$

Hence $4 + 3i$ added to $5 - 2i$ is $9 + i$. As in previous sections, we can refer to this justification as "combining like terms", and see that adding complex numbers just comes down to combining like terms. Similarly, we can multiply these two numbers together as follows:

$$(4 + 3i) \cdot (5 - 2i)$$

$=$	$4 \cdot (5 - 2i) + 3i \cdot (5 - 2i)$	Distributive Law
$=$	$(4 \cdot 5) - (4 \cdot 2i) + (3i \cdot 5) - (3i \cdot 2i)$	Distributive Law; Distributive Law
$=$	$20 - 8i + 15i - 6 \cdot i \cdot i$	Associative and Commutative Laws of Multiplication
$=$	$20 - 8i + 15i - 6 \cdot (^-1)$	Definition of the number i
$=$	$20 - 8i + 15i + 6$	$^-6 \cdot {^-1} = 6$
$=$	$26 + 7i$	"Combining Like Terms", as previously justified

Homework Set 10

1) Make a Venn Diagram showing the relationship between the sets of numbers defined in this section: Natural Numbers, Integers, Rational Numbers, Real Numbers, Complex Numbers. Where are the irrational numbers in your diagram? Where are the imaginary numbers?

2) Give two examples of each of the following
   a) An integer that is a natural number
   b) An integer that is not a natural number
   c) A rational number that is an integer
   d) A rational number that is not an integer
   e) A real number that is rational
   f) A real number that is not rational
   g) A complex number that is real
   h) A complex number that is not real.

3) Approximate the following (valid to two decimal places. You can use a calculator, but pretend that any root or exponent buttons you have do not work.
   a) $\sqrt{10}$
   b) $\sqrt[3]{9}$

4) Consider the equation $x^2 = c$. Find an integer value for c such that:

   a) The equation $x^2 = c$ does not have a solution in the real numbers, but does have a solution in the complex numbers.

   b) The equation $x^2 = c$ does not have a solution in the rational numbers, but does have a solution in the real numbers.

   c) The equation $x^2 = c$ does not have a solution in the integers, but does have a solution in the rational numbers.

5) Find all values for x that make the following equations true. Assume x is a complex number.

   a) $x^2 + 4 = 0$
   b) $5 + x^2 = 2$
   c) $x^3 + 8 = 0$
   d) $x^4 - 81 = 0$

6) Write the following numbers in the form $a + bi$, where a and b are real.

   a) i^2
   b) i^3
   c) i^4
   d) i^5
   e) i^{10}
   f) i^{100}

7) Write the following numbers in the form $a + bi$, where a and b are real.
   a) $(1 + i)^2$
   b) $i \cdot (3 + 2i) - 3(2 + 5i)$

8) Figure out what complex number is i^{-1}. In other words, what number is the multiplicative inverse of i?

9) Prove that $\frac{1}{2} - \frac{1}{2}i$ is the multiplicative inverse of $1 + i$.

10) Assume that we can apply the definition of square roots to complex numbers as well. Prove that $\sqrt{i} = \frac{1}{\sqrt{2}} + \frac{1}{\sqrt{2}}i$.

11) Given a complex number $a + bi$, the number $a - bi$ is called its **complex conjugate**. Show that $(a + bi) \cdot (a - bi)$ is always a real number.

12) A technique that works well for simplifying a division by a complex number is to multiply by the "complex conjugate" so that the divisor becomes real. (See the previous problem).

Here's some examples:

a) Consider simplifying $\frac{1}{3+5i}$ by first multiply both numerator and denominator by the complex conjugate of $3 + 5i$. We'll get you started, and you finish.

$$\frac{1}{3+5i}$$
$$= \frac{1}{3+5i} \cdot 1 \qquad \text{Definition of Multiplicative Identity}$$
$$= \frac{1}{3+5i} \cdot \frac{3-5i}{3-5i} \qquad \text{Definition of Multiplicative Inverses}$$
$$= \frac{1 \cdot (3-5i)}{(3+5i) \cdot (3-5i)} \qquad \text{Fraction Multiplication}$$
$$=$$
$$=$$

b) Another way to interpret the expression $\frac{1}{3+5i}$ is that it's $(3+5i)^{-1}$, the multiplicative inverse of $3 + 5i$. Show that the number you found in part a) is indeed the multiplicative inverse of $3 + 5i$.

c) Use the technique in part a) to simplify $\frac{3+i}{1-2i}$.

Class Activity 11: Oversimplification

We like to test things ... no matter how good an idea sounds, test it first.

Henry Block

Determine which of the following are valid algebraic simplifications.

- If the two expressions are equivalent (have the same value for relevant values of x), prove it using definitions, laws and properties we have discussed in the text.

- If the two expressions are not equivalent, prove it by giving two examples of choices for the value of x which show the expressions are not equivalent.

1. Is $\sqrt{16 + 4x^2}$ equivalent to $4 + 2x$?

2. Is $(x - 2)^5$ equivalent to $x^5 - 32$?

3. Is $(x + 3)^4 - 5(x + 3)^4$ equivalent to $^-4(x + 3)^4$?

4. Is $\frac{x^2+7}{x}$ equivalent to $x + 7$?

5. Is $\frac{2x^2-x}{x}$ equivalent to $2x - 1$?

6. Is $5(x - 1)^2$ equivalent to $(5x - 5)^2$?

7. Is $\frac{1}{x+2}$ equivalent to $\frac{1}{x} + \frac{1}{2}$?

Chapter Two

Equations

Class Activity 12: Equality for All!

1. Which value(s) for x will make the following equations true? Which value(s) for x will make the equation false?

    a. $3(x - 2) = 2x - 1$

    b. $3(x - 2) = 3x - 6$

    c. $3(x + 1) = 3x - 2$

2. Imagine two giant balance scales shown below. Two donkeys and a bear weigh the same as an elephant, and an elephant and a donkey weight the same as three bears.

   a) How many donkeys does it take to balance a bear? First try to do the problem without writing down any equations and explain how you thought about it.

   b) Now translate the problem to a system of equations and solve using algebraic methods.

   c) If an elephant weighs 3000 pounds, what does a bear weigh? A donkey?

85

Read and Study 12: Types of Equations

Equations are the devil's sentences.

Stephen Colbert

In algebra, an **equation** is a statement asserting that two expressions have the same numerical value. An **inequality** is a statement that one expression has a value that is less or greater than another expression.

Like all statements, equations and inequalities can by true or false. For example, the statement "Madison is the capital of Wisconsin" is a true statement, while "Oshkosh is the capital of Wisconsin" is a false statement. All equations and inequalities fall into one of three categories: identity, contradiction, or conditional, depending on whether they are always true, never true, or sometimes true, respectively.

An **identity** is an equation or inequality that is true for values of the variable that we care about. For example, the equation

$$2x - 3 = 1 + x - 4 + x$$

is always true no matter what value x takes, and the inequality

$$2t < 2t + 3$$

is true no matter what value t takes. The equation

$$\frac{5x(x-3)}{(x-3)} = 5x$$

would also be called an identity, because it is true for all values of x except for one exception where the expression on the left side is not defined. *What is that exception?*

A **contradiction** is an equation or inequality that is *never* true, no matter what values the variables take. Examples here include:

$$3x - 4 = 3x + 7, \text{ and}$$

$$a - 1 > a + 3.$$

Take a moment to convince yourself that those equations are never true no matter what the values for x or a may be.

A **conditional** equation or inequality is true for some real numbers, and false for others. For example, the equation

$$2x + 1 = 7$$

is true when x is 3, and false if x is any other number. Similarly, the inequality

$$w - 3 \leq 4$$

is true if w is 7 or any number less than 7, and false if w is greater than 7.

Looking back at the previous chapter, you can see it is full of identities. Here are just a few:

$$a(bc) = (ab)c$$

$$a(b + c) = ab + ac$$

$$\frac{a}{b} \cdot \frac{c}{d} = \frac{ac}{bd}$$

$$a^n a^m = a^{n+m}$$

$$\sqrt[n]{ab} = \sqrt[n]{a}\sqrt[n]{b}$$

These equations are all identities, since we were claiming that they are true for all stated possible values for a, b, n, et cetera. A **solution** to an equation (or inequality) is a value for the variable or variables that makes the equation (or inequality) true. For an identity, all possible choices for the variables are solutions. For a contradiction, no possible choices for the variables are solutions. For a conditional equation, some choices for the variables are solutions, and some are not. **Solving an equation** or inequality means finding all the values of the variables, if any exist, which make the given equation or inequality true.

To continue with our earlier analogy, suppose we had a statement: "X is the capital of Wisconsin", where the variable X represents a city. Then this statement would be a conditional statement, in that its truth is conditional on the value of X. The solution to this statement would Madison, since the statement is true when X is Madison, and it is false when X is Oshkosh, or any other city.

Please note that the word 'solution' has a special meaning with regards to equations that is much more specific than the general use of the word. In general English, a 'solution' is a way to fix a problem. And even in math, if you are not talking about an equation, the word 'solution' is often used to refer to the 'answer' or more accurately 'a method for finding the answer' to a problem. (Some math texts have a 'solutions' manual, where they give the steps you can take to 'solve' the problem). But in the context of equations, the 'solution' to an equation does not mean 'the answer' or 'a method for finding an answer', but rather it means 'the values for the variables that make the equation true'. And 'solving an equation' does not refer to a specific process or sequence of steps, but rather refers to the act of figuring out what is and what isn't a solution.

For example, if we're asked to solve the conditional equation

$$1 - 4(x - 3) = {}^-2x + 3(x + 1),$$

that means we are being asked to find the values for x that make this equation true. We could try to see if $x = 0$ is a solution, and we would find that if x were 0, then the equation would become $13 = 3$, which is false, so we know that x being 0 is not a solution. If we investigated further, and kept on substituting value after value for x, we would find that there is only one real number that makes this equation true, namely the number 2. So x being 2 is "the" solution to this equation.

Note that if for some reason when we were first presented with this problem we had suspected that $x = 2$ might be a solution, and then we verified that it was by substituting the value into the equation to see that indeed worked (do this!), then we would have 'solved the equation'. However, a "guess and check" method is usually not a very efficient way to find solutions, and more importantly, just because we may know that 2 is a solution, how do we know that there aren't any other solutions? We will discuss these questions further in the next section.

In the Donkey, Elephant and Bear problem in the Class Activity, you saw a way that algebraic equations can be thought of as balance where the expressions on either side of the equals sign represent weights, and that those weights are in balance. With this view, to find a solution to the equation (or system of equations) we can use techniques such as substitute equal weights to each other, grouping equal weights together, and removing equal weights from both sides of the balance. We will make explicit connections between these techniques and the Laws of Algebra in the next section. For now, we will use your intuition on keeping equal weights balanced to find solutions to equations like these.

Having pictures or actual objects representing the weights that we can manipulate is a great way to solve these balance equation problems. Commercially available sets of tiles (e.g., Alge-tiles™ or Algebra Tiles™, or Algebra Models™) can be used to represent variable quantities. In this section, we will use red squares representing the unit constant, a green rectangle representing a variable x and an orange rectangle representing a variable y. You can print and cut out your own set using the template in the appendix.

The small red squares have side lengths of 1, and so have an area of 1.
The green rectangles have side lengths 1 and x and so have an area of x.
The orange rectangles have side lengths 1 and y, and so have an area of y.

For example, the equation $4x + 3 = 2x + 7$ could be represented with the balance shown on the left below. Four weights with value x and three weights with value 1 on the left side of the balance, have the same weight as two weights of value x and 7 weights with value 1.

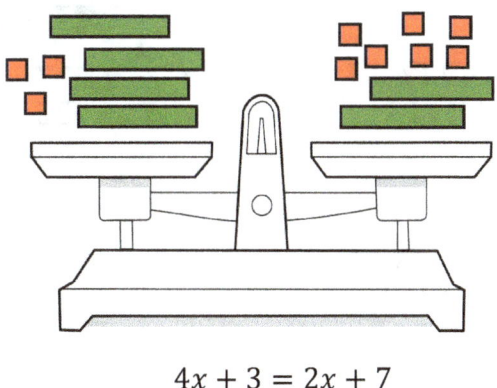

$$4x + 3 = 2x + 7$$

"Solving" this equation means finding the value for the variable x. In this context, we want to know how many red "ones" weights it takes to balance one green "x" weight. We can simply things first by removing two "x" weights and three "ones" weights from each side. Since we have removed the same weight from each side, the equation will still balance.

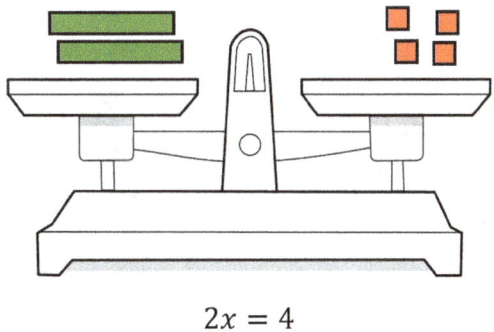

$$2x = 4$$

Now we see that two "x" weights balance four "ones" weights. Then we can divide the weights into groups to see that one green x weight must balance with 2 red "ones" weights.

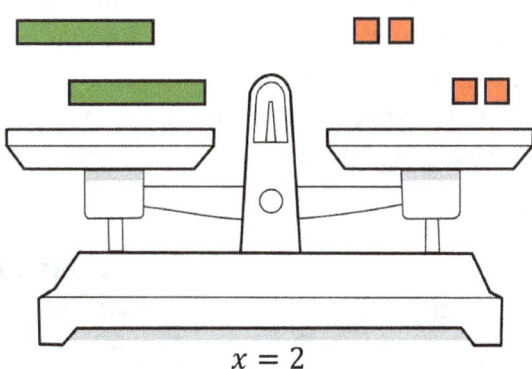

$$x = 2$$

A system of two linear equations can also be solved by thinking about two balance scales. For example, suppose we have the two equations $x + 2y = 6$ and $3x + 2y = 12$. This could be represented by these two balances:

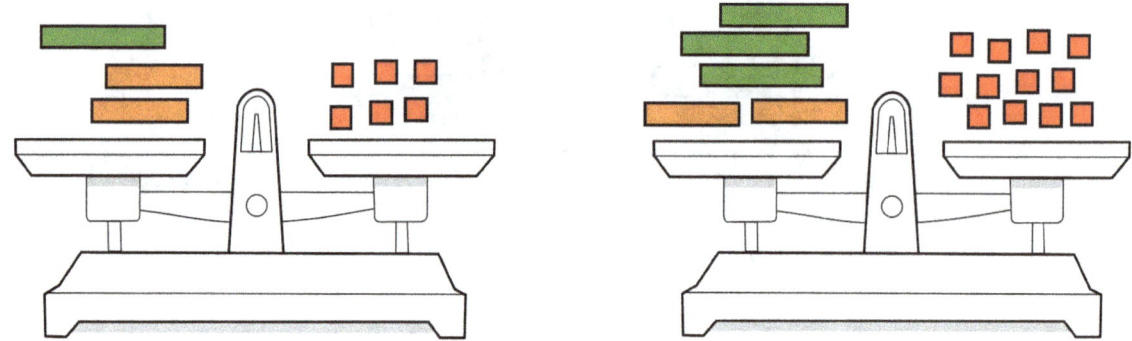

Since each equation has two different variables, we don't know how much an x or y weighs just by looking at just one balance or the other. And there's nothing obvious we could do to simplify *either* balance. But we could substitute or remove equal weights. For example, the first balance shows that a green and two orange weights is the same as 6 reds. So on the second balance, we could remove a green and two orange tiles from the left side, and remove 6 reds from the right. Since we have removed equal weights from each side, the new equation will still balance.

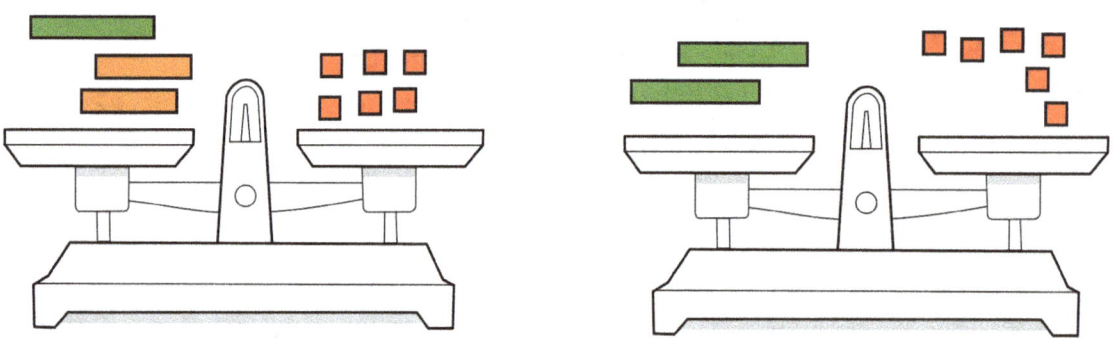

Now we can group the tiles on the second balance to see that one green x tile must weigh the same as three red "ones", in other words, that $x = 3$. To find out the value of y, we could then use the first balance. On this balance, let's remove one green x from the left side and three red "ones" (since we already figured out that $x = 3$, we are removing the same weight from both sides of the balance.) That leaves us with two orange y's balancing three red "one"s. So each y must be one and a half ones, or $y = \frac{3}{2}$.

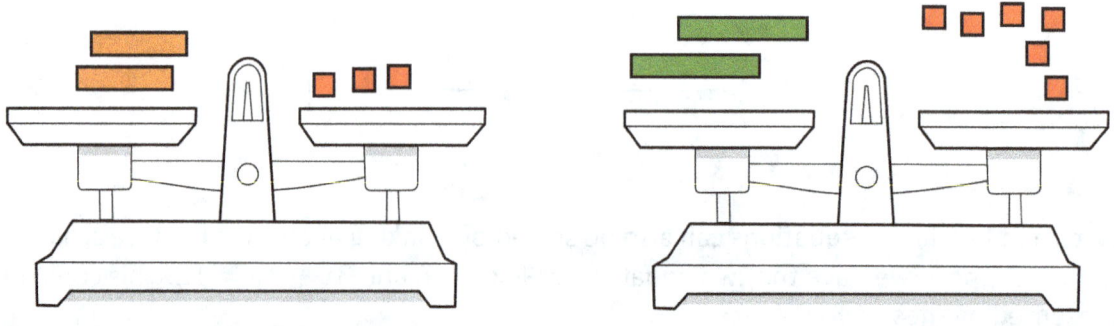

So we have found that for the equations to be true, x must have a value of 3 and y must have a value of $\frac{3}{2}$. Take a moment to check that these values make both original equations $x + 2y = 6$ and $3x + 2y = 12$ true.

Homework Set 12

1) Write down the sequence of algebraic equations that corresponds to each step of how we solved the equation $4x + 3 = 2x + 7$ using the balance in the Read and Study. For each new equation, write down what was done to the previous equation to get the new equation.

2) Write down the sequence of algebraic equations that corresponds to each step of how we solved the system of equations $x + 2y = 6$ and $3x + 2y = 12$ using the balance in the Read and Study. For each new equation, write down what was done algebraically to the previous equation to get the new equation.

3) Determine whether the following equations are identities, contradictions, or conditional. Justify your answer.

   a) $3(x-1)^2 = 3(x^2 - 1)$

   b) $3(x-1)^2 = 3x^2 - 6x + 3$

   c) $3(x-1)^2 = 0$

   d) $\frac{5}{x} = \frac{1}{2}$

   e) $\frac{5}{x} = \frac{3}{x}$

   f) $\frac{6}{2x} = \frac{3}{x}$

4) Classify each of the following equations as an identity, contradiction, or conditional:

   a) $\sqrt{16 + 4x^2} = 4 + 2x$

   b) $(x-2)^5 = x^5 - 32$

   c) $(x+3)^4 - 5(x+3)^4 = {^-4}(x+3)^4$

   d) $\frac{x^2+7}{x} = x + 7$

   e) $\frac{5(x-1)^2}{x-1} = 5(x-1)$

   f) $\left(\frac{1}{x} + \frac{1}{2}\right)^{-1} = x + 2$

   g) $5(x-1)^2 = (5x-5)^2$

91

5) If possible, find a value for the number c so that the equation

$$5x = c(x + 4)$$

   a) Is a conditional equation with $x = 6$ as a solution.

   b) Is a contradiction or an identity

6) If possible, find a value for the number c so that the equation

$$cx + 5 = 3x + 2c - 1$$

   a) Is a conditional equation with $x = 2$ as a solution.

   b) Is a contradiction or an identity

7) Show how you can use Algebra tiles to solve the following equations.

   a) $3(x + 4) = 4x + 7$

   b) $5x + 2 = 3x + 10$

8) Show how you can use Algebra tiles to solve the following systems of two equations.

   a) $x + y = 5$
      $2x + 3y = 7$

   b) $2x + y = 3y + 3$
      $3x + 4y = 2x + 4$

9) Consider the following scenario. Three chickens weigh the same as monkey. Two monkeys and a chicken weigh the same as a sheep.
   a) Using the variables C, M, and S, write two equations that represent the relationships in the problem.
   b) Carefully define what each of the variables C, M, and S, represent in your equations.
   c) If a sheep weighs 35 pounds, how much does a monkey weigh?

Class Activity 13a: Follow the Law

The laws of addition, multiplication, and equality are exactly the ones we need to make our algebra work the way it does. Suppose we have the equation $7(x + 3) = 10x$ and we want to solve for x. Justify that each subsequent equation in the series of equations below follows from the previous equation. Cite the relevant definition, property, or theorem from the text

$$7 \cdot (x + 3) = 10 \cdot x$$

$$(7 \cdot x) + (7 \cdot 3) = 10 \cdot x$$

$$(7 \cdot x) + 21 = 10 \cdot x$$

$$^-(7 \cdot x) + [(7 \cdot x) + 21] = {}^-(7 \cdot x) + (10 \cdot x)$$

$$[^-(7 \cdot x) + (7 \cdot x)] + 21 = {}^-(7 \cdot x) + (10 \cdot x)$$

$$0 + 21 = {}^-(7 \cdot x) + (10 \cdot x)$$

$$21 = {}^-(7 \cdot x) + (10 \cdot x)$$

$$21 = (^-7 \cdot x) + (10 \cdot x)$$

$$21 = (^-7 + 10) \cdot x$$

$$21 = 3 \cdot x$$

$$\tfrac{1}{3} \cdot 21 = \tfrac{1}{3} \cdot (3 \cdot x)$$

$$7 = \tfrac{1}{3} \cdot (3 \cdot x)$$

$$7 = \left(\tfrac{1}{3} \cdot 3\right) \cdot x$$

$$7 = 1 \cdot x$$

Class Activity 13b: Equivalent or Not Equivalent?

Alice, Bob and Charlie were each working on solving the equation $(x + 3)^2 = 4x^2 - 36$ by writing a sequence of equivalent equations. Below is the next equation that each student wrote. For each, decide what the student did, and whether that results in an equivalent equation. Explain why or why not.

    a) Alice: $x + 3 = 2x - 6$

    b) Bob: $x^2 + 9 = 4x^2 - 36$

    c) Charlie: $x^2 = 4x^2 - 39$

David, Elise, and Francis were each working on solving the equation: $x + \sqrt{x^2 + 9} = 4$ by writing a sequence of equivalent equations. Below is the next equation that each student wrote. For each, decide what the student did, and whether that results in an equivalent equation. Explain why or why not.

    d) David: $x^2 + (x^2 + 9) = 16$

    e) Elise: $x^2 + 9 = \sqrt{4 - x}$

    f) Francis: $x + (x + 3) = 4$

Gary, Hannah, and Isabelle were each working on solving the equation: $\frac{3x}{x+4} = 2x - 5$ by writing a sequence of equivalent equations. Below is the next equation that each student wrote. For each, decide what the student did, and whether that results in an equivalent equation. Explain why or why not.

    g) Gary: $\frac{3}{4} = 2x - 5$

    h) Hannah: $3x = 2x - 5(x + 4)$

    i) Isabelle: $\frac{x}{x+4} = -5$

Read and Study 13: Properties of Equality and Solving Equations

Love and you shall be loved. All love is mathematically just, as much as the two sides of an algebraic equation.

Ralph Waldo Emerson

At the end of the previous section, we considered the problem of finding solutions to the conditional equation

$$1 - 4(x - 3) = {}^-2x + 3(x + 1).$$

We noted that $x = 0$ is not a solution to the equation, while $x = 2$ is a solution. But how could we have found the solution $x = 2$ without guessing and checking? And how do we know whether $x = 2$ is the *only* solution?

The algebraic method is to use power of logic to argue that the given equation

$$1 - 4(x - 3) = {}^-2x + 3(x + 1),$$

and the equation

$$x = 2$$

are equivalent equations. Two equations or inequalities are called **equivalent** if they have exactly the same solutions. Since obviously the equation $x = 2$ is true only when x takes the value of 2, if we can show that our original equation is equivalent to $x = 2$, we know that the original equation is true only when x is 2.

Properties of Equality
(Procedures which Generate Equivalent Equations)

The following do not change the solutions to an equation:

- Adding (or subtracting) the same real number to (from) both sides of the equation.

- Multiplying (or dividing) both sides of the equation by the same **nonzero** real number.

> **Properties of Inequality**
> (Procedures which Generate Equivalent Inequalities)
>
> The following do not change the solutions to an inequality:
>
> • Adding (or subtracting) the same real number to (from) both sides of the inequality.
>
> • Multiplying (or dividing) both sides of the inequality by the same **positive** real number.
>
> If you multiply both sides of an inequality by a negative real number, the inequality sign is reversed. For example: $3 < 4$, but if you multiply both sides by $^-2$, you get $^-6 > ^-8$.

The big idea in solving a equation is to use the definitions and properties of arithmetic to re-write the equation as an equivalent equation, then rewrite that one into another equivalent equation, and so on, until we arrive an equivalent equation where the solution is obvious.

Suppose we didn't know that $x = 2$ was a solution to $1 - 4(x - 3) = {^-2x} + 3(x + 1)$. What follows is a possible sequence of equivalent equations that could be used to find that solution. Proficient equation solvers would likely take shortcuts that combine several of these steps, but our goal in writing this out is not to show you how to get the answer as quickly as possible, but to point out the various properties and definitions that justify that each equation is equivalent.

$$1 - 4(x - 3) = {^-2x} + 3(x + 1)$$

$$1 + {^-4}(x + {^-3}) = {^-2x} + 3(x + 1) \quad \text{Definition of subtraction}$$

$$1 + ({^-4x} + 12) = {^-2x} + 3(x + 1) \quad \text{Distributive Law}$$

$$1 + ({^-4x} + 12) = {^-2x} + (3x + 3) \quad \text{Distributive Law}$$

$$1 + (12 + {^-4x}) = {^-2x} + (3x + 3) \quad \text{Commutative Law of Addition}$$

$$(1 + 12) + {^-4x} = ({^-2x} + 3x) + 3 \quad \text{Associative Law of Addition}$$

$$13 + {^-4x} = x + 3 \quad \text{'Combine Like Terms'}$$

$${^-3} + (13 + {^-4x}) = (x + 3) + {^-3} \quad \text{Add } {^-3} \text{ to both sides}$$

$$({^-3} + 13) + {^-4x} = x + (3 + {^-3}) \quad \text{Associative Law of Addition}$$

$$10 + {}^-4x = x + 0 \qquad \text{Additive Inverses}$$

$$10 + {}^-4x = x \qquad \text{Additive Identity}$$

$$(10 + {}^-4x) + 4x = x + 4x \qquad \text{Add } 4x \text{ to both sides}$$

$$10 = 5x \qquad \text{Associative Law; 'Combine Like Terms'}$$

$$\frac{10}{5} = \frac{5x}{5} \qquad \text{Multiply both sides by } 5^{-1}$$

$$2 = x$$

Wow. That's thorough. The big point here is every, and we mean *every* thing we do in algebra works because it is justified by one of the definitions and laws from the last chapter. Notice that are lots of times when we repeatedly do the same sequence of steps, and justifying each step with a definition or law every time becomes very tedious. That's why we have come up with names for certain packages of procedures like **'combining like terms'** (which uses Associativity, Commutativity, and the Distributive Law) and **'canceling'**, (which uses Associativity, Commutativity, Inverse and Identity Laws) which we have already justified in the last chapter, to justify combined steps.

For example, in the last chapter, we justified the last step above as follows:

$$\frac{5x}{5} = (5x) \cdot \frac{1}{5} \qquad \text{Definition of Division}$$

$$= \frac{1}{5} \cdot (5x) \qquad \text{Commutative Law of Multiplication}$$

$$= \left(\frac{1}{5} \cdot 5\right) x \qquad \text{Associative Law of Multiplication}$$

$$= 1 \cdot x \qquad \text{Definition of Multiplicative Inverses}$$

$$= x \qquad \text{Definition of Multiplicative Identity}$$

In practice, many of the steps would be combined, but it's crucial to keep in mind the justification for why each subsequent is equivalent to the previous equation. What follows is a typical condensed version with lots of combined steps:

$$1 - 4(x - 3) = {}^-2x + 3(x + 1)$$

$$1 - 4x + 12 = {}^-2x + 3x + 3 \quad \text{Definition of Subtraction, Distributive Law}$$

$$13 - 4x = x + 3 \quad \text{Commutative and Associative Laws; 'Combine Like Terms'}$$

$$10 - 4x = x \quad \text{Subtract 3 from both sides; Associative Law}$$

$$10 = 5x \quad \text{Add } 4x \text{ to both sides, Associative Law; 'Combine Like Terms'}$$

$$2 = x \quad \text{Divide both Sides by 5.}$$

You've likely been told before you should always check your answer by substituting $x = 2$ into the original equation. If it's important to know you have the right answer, this is a good idea, just in case you've made some small error along the way. But mathematically, it is not necessary to check whether 2 is a solution. That's because we've justified that each of the equations above is equivalent to every other. So they all have the same solutions. Just as we know that 2 a solution (and in fact, the *only* solution) to the equation $x = 2$, we know that 2 is the only solution to any of the earlier equations, including the one we started with.

In the box above for generating equivalent equations, we said that you can multiply (or divide) both sides of the equation by the same **nonzero** real number. In the homework we will ask you to think about why we need to say that the number cannot be zero. Now it would likely not occur to you that you would want to multiply both sides of an equation by zero, or divide both sides of an equation by zero, but you might often be in the situation where you'd like to multiply or divide both sides of an equation by a variable (or an expression involving a variable). Whenever you do this, you need to stipulate that you recognize that the resulting equation is no longer equivalent to the original if the quantity you multiplied by or divided by is zero. Let's explore an example where this issue arises.

Suppose you are given an equation in two variables, such as

$$yx - 1 = 3y + x,$$

and for some reason we want to solve this equation for y, meaning we want to write this as an equivalent equation with y isolated on one side of the equation. First we notice that there is a yx term on the left side, and a $3y$ term on the right side. So let's first get the expressions with a y together on one side of the equation. Here's what that would look like:

$$yx - 1 = 3y + x$$

$$yx - 1 + 1 = 3y + x + 1 \quad \text{Add 1 to both sides}$$

$$yx = 3y + x + 1 \qquad \text{Additive Inverses; Additive Identity}$$

$$yx - 3y = 3y + x + 1 - 3y \qquad \text{Add } ^-3y \text{ to both sides}$$

$$yx - 3y = 3y - 3y + x + 1 \qquad \text{Commutativity of Addition}$$

$$yx - 3y = x + 1 \qquad \text{Additive Inverses; Additive Identity}$$

$$y \cdot (x - 3) = x + 1 \qquad \text{Distributive Law; Additive Identity}$$

$$y \cdot (x - 3) \cdot \frac{1}{(x-3)} = (x+1) \cdot \frac{1}{(x-3)} \qquad \text{Multiply both sides by } \tfrac{1}{x-3}\text{, provided } x \neq 3$$

$$y \cdot 1 = (x+1) \cdot \frac{1}{(x-3)} \qquad \text{Multiplicative Inverses}$$

$$y = \frac{x+1}{x-3} \qquad \text{Multiplicative Identity; Def. of Division}$$

Notice how in our third to the last step we multiplied by $\frac{1}{x-3}$. In order for this to be justified, we need to be sure that $(x - 3)$ is not zero, since zero doesn't have a multiplicative inverse. (Recall our previous discussion about why "you can't divide by zero".) But $(x - 3)$ *would* be zero if x were 3. So we need to stipulate that $x \neq 3$ in order for us to say the equations are equivalent.

The result of our analysis is that we can say that the equations

$$yx - 1 = 3y + x,$$

and

$$y = \frac{x+1}{x-3}$$

are equivalent, **provided $x \neq 3$**.

That means these two equations have the same solutions, provided we ignore the case where x is 3. Well 3 is only one number out of the (uncountably) infinite number of real numbers, so that's not too bad! But it does mean that it's possible that 3 could *potentially* be a solution to one of the equations, but not the other.

So out of curiosity, what would each equation look like if x were 3? The original equation would become $3y - 1 = 3y + 3$, which is a contradiction. The second equation would simplify to $y = \frac{4}{0}$, which is undefined. So in either case, both equations yield a contradiction when $x = 3$, so 3 is not a solution to either equation. So it turns out in this case it wouldn't have been a big deal if we had ignored that we might be dividing by zero. But as we will see in later examples, it can be a big deal, a very big deal. That's why we are paying so much attention right from the start to the logic of solving equations.

In our experience, many students struggle with equations like the one in our previous example, and try to solve for y by "canceling" x in ways that violate the laws of algebra. For example, you might be tempted to try to get rid of the x in that yx term dividing by x. But in order to do this, you would have to divide **both sides** of the equation by x, which means you'd have to distribute $\frac{1}{x}$ to each term on both sides of the equation (not just selectively to the ones you want to!). We'll show you what this would have to look like:

$$yx - 1 = 3y + x$$

$$\frac{1}{x} \cdot (yx - 1) = \frac{1}{x} \cdot (3y + x) \quad \text{Multiply both sides by } \frac{1}{x}\text{; provided } x \neq 0.$$

$$yx \cdot \frac{1}{x} - \frac{1}{x} = 3y \cdot \frac{1}{x} + x \cdot \frac{1}{x} \quad \text{Distributive Law}$$

$$y - \frac{1}{x} = \frac{3y}{x} + 1 \quad \text{Multiplicative inverses; Multiplicative Identity}$$

So while dividing by x does indeed get the y all alone in that first term on the left side, we also have a $\frac{-1}{x}$ term as well. But more inconveniently we have a $\frac{3y}{x}$ term on the right side. And so we still would need to subtract that $\frac{3y}{x}$ and use the distributive law to factor out a y in order to isolate the y. So dividing by x right away in this example turns out to not really be helpful.

Homework Set 13

1) A student was given the equation $3(x - 2) = 5x + 8$ and was asked to solve the equation. Here are the steps the student wrote. Justify each step by citing the appropriate definition or property or Law of Algebra used.

$$3(x - 2) = 5x + 8$$

$$3x - 6 = 5x + 8$$

$$3x - 6 + 6 = 5x + 8 + 6$$

$$3x + {}^-6 + 6 = 5x + 14$$

$$3x + 0 = 5x + 14$$

$$3x = 5x + 14$$

$$3x + {}^-5x = 5x + 14 + {}^-5x$$

$(3 + {}^-5)x = 5x + 14 + {}^-5x$

${}^-2x = 5x + 14 + {}^-5x$

${}^-2x = 5x + {}^-5x + 14$

${}^-2x = 14$

${}^-2x \cdot \frac{-1}{2} = 14 \cdot \frac{-1}{2}$

${}^-2 \cdot \frac{-1}{2} \cdot x = 14 \cdot \frac{-1}{2}$

$1 \cdot x = 14 \cdot \frac{-1}{2}$

$x = {}^-7$

2) Solve the following equation or inequality. You can combine steps together, but you must state the appropriate definition or property or Law of Algebra used that justify that each new equation or inequality is equivalent to the previous one. Confirm that the solutions to your last equation are indeed solutions to the first equation or inequality by substituting those values for x into the original one.

   a. $x + 3 = 5 - 2(x + 3)$

   b. $\frac{3x-1}{5} + 1 > 4 - x$

   c. $\frac{3x+1}{2} < 4x - 3$

3) In chemistry, the if a volume V of a gas is at pressure P and temperature T, when volume, pressure and/or temperature changes, the new volume V', new pressure P' and new temperature T' are related by the equation

$$V' = V \cdot \frac{P}{P'} \cdot \frac{T'}{T}$$

   a. Solve the gas equation for T'.
   b. Solve the gas equation for P'.

4) Solve the equation $yx - 1 = 3y + x$ for x. by writing a sequence of equivalent equations until you have an equation with x isolated on one side of the equation. Justify each equation is equivalent to the previous with a definition, property or Law of Algebra. (Note: In the Read and Study, we discussed how we could solve this equation for y. Now we are asking you to solve it for x.)

5) Solve the equation $w \cdot (3 - v) = 5w - 1$ for w by writing a sequence of equivalent equations until you have an equation with w isolated on one side of the equation. Justify each equation is equivalent to the previous with a definition, property or Law of Algebra.

6) Solve the equation $w \cdot (3 - v) = 5w - 1$ for v by writing a sequence of equivalent equations until you have an equation with v isolated on one side of the equation. Justify each equation is equivalent to the previous with a definition, property or Law of Algebra.

7) In generating equivalent equations, we said that you can multiply (or divide) both sides of the equation by the same **nonzero** real number. Why? What happens to the set of solutions to a conditional equation if you multiply both sides of the equation by zero? What happens to the set of solutions to a conditional equation if you divide both sides of the equation by zero?

8) In problems 5, 6 or 7 above, did you ever multiply or divide by a quantity that could be zero? If so, go back and stipulate for which values of the variable your results hold.

9) The absolute value of a number is the distance that number is from zero on the number line.
    a. Use a complete sentence to explain what the equation $|4x + 2| = 2$ means in words.
    b. Draw a number line where the distance between two adjacent integers is roughly 1 inch. Then plot all the solutions to the equation $|4x + 2| = 2$ on the number line.
    c. Use a complete sentence to explain what the inequality $|4x + 2| > 2$ means in words.
    d. Highlight all the solutions to the inequality $|4x + 2| > 2$ on the number line.

Class Activity 14a: Let's Get Coordinated

The **graph** of an equation in two variables is the set of all ordered pairs that make the equation true. For each equation below, give two examples of ordered pairs that make the equation true, and two examples of ordered pairs that do not make the equation true. Then draw a graph of the equation. (Assume that x and y represent real numbers).

A) $3x + 4y = 12$

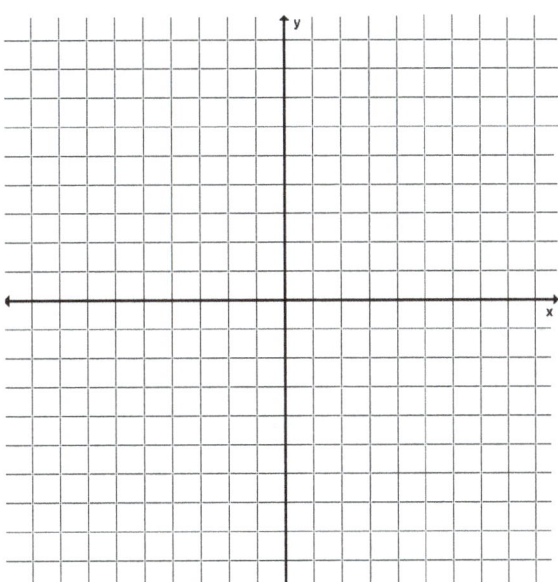

B) $x^2 + y^3 = 1$

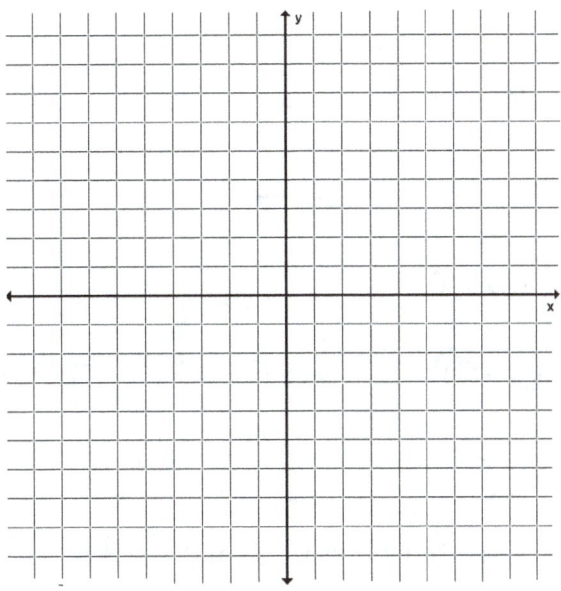

Class Activity 14b: How Do You Solve an Equation Like Maria?

Maria was given some equations to solve. Here is her work.
Your job is to justify her steps and to explain and fix any mistakes with Maria's solutions.

1. Solve for x: $\quad \frac{4x}{x+2} = 3 - \frac{8}{x+2}$.

$$4x = 3(x+2) - 8$$

$$4x = 3x+6 - 8$$

$$4x = 3x-2$$

$$x = {}^-2$$

The solution is $x = {}^-2$.

2. Solve for x: $\quad \frac{1}{3}x - 2 = \sqrt{4-x}$.

$$\left(\tfrac{1}{3}x - 2\right)^2 = 4 - x$$

$$\tfrac{1}{9}x^2 - \tfrac{4}{3}x + 4 = 4 - x$$

$$x^2 - 12x + 36 = 36 - 9x$$

$$x^2 - 3x = 0$$

$$x(x-3) = 0$$

$$x = 0 \text{ or } x = 3$$

The solutions are $x = 0$ and $x = 3$.

Read and Study 14: Techniques for Solving Equations

Logic, like whiskey, loses its beneficial effect when taken in too large quantities.
Lord Dunsany

In the previous section, we discussed how "solving an equation" means finding values for the variable(s) that make the equation true, and that a general method for finding these solutions is to continue to rewrite the equation as equivalent equations until we have an equivalent equation where the solutions(s) are obvious. Our key tool in writing equivalent equations are the Properties of Equality, which give us guarantees for when the equations we write will be equivalent, meaning they have the same solutions. We will now introduce another important property that can be very useful in finding solutions to equations.

Zero Product Property

Suppose a and b are real numbers. If one or both of a and b is zero, then the product will be zero. And if the product of a and b is zero, then at least one of the numbers a or b must be zero. In other words

$$ab = 0 \quad \text{is equivalent to} \quad a = 0 \text{ or } b = 0$$

The Zero Product Property states not only that when you multiply any number by zero, the result is zero, it also says that the *only* way to get an answer of zero when multiplying two real numbers is to have one (or both) of the numbers be zero in the first place. Hopefully, this property makes sense and you will believe it just based on your familiarity with multiplication. However, we will offer a quick proof of this property, because we want to be true to our word that everything in algebra can be derived from the definitions and Laws of Algebra by using logical reasoning.

Let a and b be real numbers where at least one of them is zero. Let's say $a = 0$. Then we'd have:

$$a \cdot b = 0 \cdot b \quad \text{We started by assuming } a \text{ is zero.}$$

$$= (c + {}^-c) \cdot b \quad \text{By definition of additive inverse, where } c \text{ is any real number.}$$

$$= c \cdot b + ({}^-c) \cdot b \quad \text{Distributive Law}$$

$$= c \cdot b + (^-1 \cdot c) \cdot b \qquad \text{By Property 1 of Additive Inverses}$$

$$= c \cdot b + {}^-1 \cdot (c \cdot b) \qquad \text{Associativity of Multiplication}$$

$$= c \cdot b + {}^-(c \cdot b) \qquad \text{By Property 1 of Additive Inverses}$$

$$= 0 \qquad \text{Definition of Additive Inverse}$$

So we've proven that if $a = 0$, then $a \cdot b = 0$. In other words, any number times zero gives you zero.

To prove the other part of the property, now let a and b be real numbers whose product is zero. Suppose the number a is *not* zero. Then by the multiplicative inverse law, a has a multiplicative inverse $\frac{1}{a}$ such that $\frac{1}{a} \cdot a = 1$. Then

$$a \cdot b = 0 \qquad \text{We started by assume the product of a and b is zero}$$

$$\frac{1}{a} \cdot a \cdot b = \frac{1}{a} \cdot 0 \qquad \text{Property of Equality: multiply both sides by } \frac{1}{a}$$

$$\frac{1}{a} \cdot a \cdot b = 0 \qquad \text{We just proved that any number times zero is zero}$$

$$1 \cdot b = 0 \qquad \text{Definition of Multiplicative Inverse}$$

$$b = 0 \qquad \text{Definition of Multiplicative Identity}$$

Now it's likely you've known what we're calling the Zero Products Property for a long time, since elementary school really, when you started learning about multiplication. But our point once again in doing these proofs is that everything in algebra makes sense and follows from the definitions and laws of Algebra.

And the Zero Products Property deserves this special attention, since this property drives many equation-solving methods in algebra, by allowing us to take complicated equations and reduce them to simpler ones. As an example, suppose we want to solve the equation $x^2 = x + 6$.

Before we start manipulating this equation, keep in mind what it means to "solve" the equation: to find all the values for the variable that make the equation true. So take a moment to think about the equation $x^2 = x + 6$. What value(s) for x would make this equation true? Remember an equation is a statement, it says something. And this equation is saying that when you square some number x, you get the same value as six more than the original number. Can you think of such a number? (Solving equations can be quite fun since they are really little puzzles to solve!)

We might reason by trial and error a little bit. Let's check what happens if x is ten. Then ten squared is 100. But that's not six more than the original number, since six more than five is only 16. We realize that squaring large numbers makes them much bigger than just adding six. Let's try a smaller number, say x is four. Then four squared is 16. That's only 12 bigger than x, but still not exactly 6 more. Then if we try $x = 3$, we've found a solution. If x is three, then x squared is 9, and six more than x is also 9.

So we now we know that $x = 3$ is a solution to the equation. But our task is to find all the solutions. Is three the only number that will work? Or could there be more? If so, how many more might there be? (Take a few minutes to see if you can find another number that works.)

The great value in our general method of writing equivalent equations is that if we can end up finding equivalent equations where the solutions are obvious, then we can be sure that we have found *all* the solutions to an equation. So let's do that now for the equation $x^2 = x + 6$. We claimed that we will try to use the zero products property. Well to do that, we need to write this equation as a product that equals zero. So we've got some work to do. We can get an equation that equals zero by subtracting the $x + 6$ from both sides, then use the distributive property to write the sum as a product.

$$x^2 = x + 6$$

$$x^2 - x - 6 = 0 \qquad \text{Property of Equality: add } {}^-x \text{ and } {}^-6 \text{ to both sides.}$$

$$(x - 3)(x + 2) = 0 \qquad \text{Distributive Property (Factoring)}$$

$$x - 3 = 0 \quad \text{or} \quad x + 2 = 0 \qquad \text{Zero Products Property}$$

$$x = 3 \quad \text{or} \quad x = {}^-2 \qquad \text{Property of Equality: add 3 to both sides;}$$
$$\text{add } {}^-2 \text{ to both sides}$$

So we've shown that the equation $x^2 = x + 6$ is logically equivalent to the statement $x = 3$ or $x = {}^-2$. And since it is obvious that the only values for x that make the statement $x = 3$ or $x = {}^-2$ true are the numbers 3 and ${}^-2$, we know that the numbers 3 and ${}^-2$ are the only values for x that make the statement $x^2 = x + 6$ true.

The great value of the Zero Products Property is that it allowed us to take a more complicated equation $x^2 - x - 6 = 0$ and write it as two much simpler equations: $x = 3$ or $x = {}^-2$.

Equations with Rational or Radical Expressions. We will close this section with some discussion of the phenomenon of "extraneous solutions" that can arise when finding solutions to equations involving rational expressions (fractions) or radicals (roots). In the Class Activity, we saw two examples of situations where we wrote equations that are not equivalent and needed to be careful about stating any provisions required for equations to be equivalent.

The first situation we had already discussed in the previous section, namely that when multiplying or dividing both sides of an equation by a quantity, we need the provision that the quantity we are multiplying by is not zero. This situation arises often when dealing with rational expressions, as it is often desirable to multiply both sides of an equation by an expression to "clear denominators", or to simplify fractions.

The second situation that arose in the Class Activity was when we are using the definition of a root to rewrite an equation. You'll notice that **there is no property of equality that says you can square both sides of an equation to get an equivalent equation**. When solving an equation with roots, we really are using the definition of the root, and requirement that even roots be positive will add a provision to our equivalent equation.

To explain further, let's consider as an example perhaps the simplest of equations involving a square root:

$$\sqrt{x} = 5$$

Recall the definition of the square root: $\sqrt{x} = y$ means that y is the non-negative number such that $(y)^2 = x$. (Read that again carefully. The definition is the most important thing to understand, since everything follows from it!) So if $\sqrt{x} = 5$, that means that $(5)^2 = x$. So x must be 25. In other words,

$$\text{If } \sqrt{x} = 5, \text{ then } x = 25.$$

These equations, $\sqrt{x} = 5$ and $x = 25$, are equivalent. They both are true only when x is 25. Now people sometimes will describe the process of going from the first equation, $\sqrt{x} = 5$, to the second equation, $x = 25$, as "squaring both sides" of the first equation, but that is **not** what's really going on. The equations are equivalent because of the definition of the square root, not because "squaring both sides" is a property of equality. In fact, "squaring both sides" is NOT a property of equality, as we will now show.

Suppose instead the equation we were to solve was

$$\sqrt{x} = {}^-5$$

Then by the definition of the square root, this equation has no solution, since $\sqrt{x}$ is always defined to be non-negative. But if we were to "square both sides", we would end up with the equation

$$x = 25$$

which obviously does have a solution. So the equations $\sqrt{x} = {}^-5$ and $x = 25$ are not equivalent, since they do not have the same solutions. So "squaring both sides" is not a property of equality.

As another example, consider the equation:

$$x = 3$$

This equation obviously has only one solution, as it's true only when x is 3. But now what it we "square both sides" of this equation? We'd end up with the following:

$$x^2 = 9$$

This equation, however, is not equivalent to the first. Why? Because this equation has TWO solutions. This equation is true when x is 3, and it's also true when x is ${}^-3$.

So this process of "squaring both sides" may not give us an equation equivalent to what we started with. It can actually generate another so-called "extraneous" solution: a number that made the new equation true, but does not make the original equation true. But if we are trying to find solutions to the original equation, we don't want this "extraneous" solution generated by "squaring both sides".

Some math texts and teachers will tell you that to solve equations with square roots, you should square both sides, and then check for extraneous solutions. But really, it's just a matter of using the definition of the square root to rewrite the equation, keeping in mind the condition in the definition that says the square root can not be negative.

Let's finish with another example of how to use the definitions and laws of algebra to solve an equation involving square roots. Consider the following equation:

$$\sqrt{2t + 3} = t - 6$$

An important skill in doing algebra is to be able to read and interpret equations. Remember an equation is a statement. It is saying something. So, pay attention to what it is saying. This equation says that the square root of the number $(2t + 3)$ is the number $(t - 6)$. By the definition, that means that $(t - 6)$ is the positive number you square to get $(2t + 3)$. So let's use that definition to rewrite the equation:

$$\sqrt{2t + 3} = t - 6$$
$$2t + 3 = (t - 6)^2 \quad \text{Definition of square root, provided } (t - 6) \geq 0$$
$$\text{In other words, } \textbf{provided } \boldsymbol{t \geq 6}$$

$$2t + 3 = t^2 - 12t + 36 \qquad \text{Distributive Property}$$

$$0 = t^2 - 14t + 33 \qquad \text{Property of Equality: Subtract } 2t + 3 \text{ from both sides}$$

$$0 = (t-3)(t-11) \qquad \text{Distributive Property (Factoring)}$$

$$(t-3) = 0 \quad \text{or} \quad (t-11) = 0 \qquad \text{Zero Products Property}$$

$$t = 3 \quad \text{or} \quad t = 11 \qquad \text{Property of Equality: add 3 to both sides in the first equation; add 11 to both sides in the second}$$

So we see that the equation $\sqrt{2t+3} = t - 6$ is equivalent to $t = 3$ or $t = 11$, provided $t \geq 6$. By the provision, t cannot be 3, which leaves only the solution $t = 11$. Since our logic is sound (you've been checking it, right?), we know that $t = 11$ is the only value for t that makes the original equation true.

But to help our understanding, or as a way to check our calculations, we can confirm this by substituting values into the original equation. We can see that 3 is indeed not solution, since it results in the equation $\sqrt{9} = {}^-3$, which is false. On the other hand, we see that 11 is in fact a solution, since it results in the equation $\sqrt{25} = 5$, which is true.

Homework Set 14

1) Can you "cancel" the x's in the following equations as shown? In other words, are the two equations equivalent? (Equivalent Equations have the same solutions). If they are, prove the two equations are equivalent using definitions and laws of algebra. If not, explain why the two equations are not equivalent.

   a) $y + 3x = 7x \quad \overset{?}{\Rightarrow} \quad y + 3 = 7$

   b) $(3 + y)x = x + 7 \quad \overset{?}{\Rightarrow} \quad 3 + y = 7$

   c) $3y + x = x + 7 \quad \overset{?}{\Rightarrow} \quad 3y = 7$

2) Maria was given another equation to solve: $x^2 + 2x - 3 = 21$.

Here work is shown below. Critique her method. Identify any invalid logical steps. Are 18 and 22 indeed solutions to the equation? Explain how you can tell. If not, what are the solutions to this equation?

$$x^2 + 2x - 3 = 21$$
$$(x + 3)(x - 1) = 21$$
$$x + 3 = 21 \text{ or } x - 1 = 21$$
$$x = 18 \text{ or } x = 22$$

3) For each equation, make an accurate sketch of its graph. Use grid paper, and show the graph with x and y between $^-10$ and 10. Be sure to find and label all integer grid points that are on the graph, and sufficient other approximate points to convey and accurate general shape of the graph.

(a) $x^2 + y = {}^-6$

(b) $x^2 + y^2 = 25$

(c) $y^2 = x^3 + 3x^2$

(d) $y = |x|$

4) For each equation given, find all the values for the variable that make the equation true. State appropriate definitions, properties or Laws of Algebra that proves each new equation you write is equivalent to the given equation, or justify with a logical argument that your solutions are the only ones.

a. $(5x + 3)^4 = 16$

b. $(2x - 5)^2 = 9$

c. $x + \sqrt{x + 2} = 4$

d. $\frac{3x}{x+4} = 2x - 5$.

e. $(x - 5)^2 = x^2 + 16$

f. $\left(\frac{1}{2}x^2 - 20\right)^3 = {}^-8$

g. $3^{(x-5)} = 81$

h. $9^{x-2} = \frac{1}{27}$

i. $\frac{x+3}{x-1} = \frac{3x-6}{x+2}$

j. $x + \frac{2}{x-5} = 4 + \frac{7-x}{x-5}$

k. $x + \sqrt{x^2 + 9} = 4$.

l. $3 - 5\sqrt[3]{2y + 1} = 0$

111

Class Activity 15: Go the Distance

1. Find the distance between the following pairs of points in the Cartesian Coordinate Plane:

    a. $(5, 3)$ and $(5, 8)$

    b. $(5, 3)$ and $(^-7, 3)$

    c. $(5, 3)$ and $(^-7, 8)$

    d. $(3, ^-2)$ and $(^-6, ^-7)$

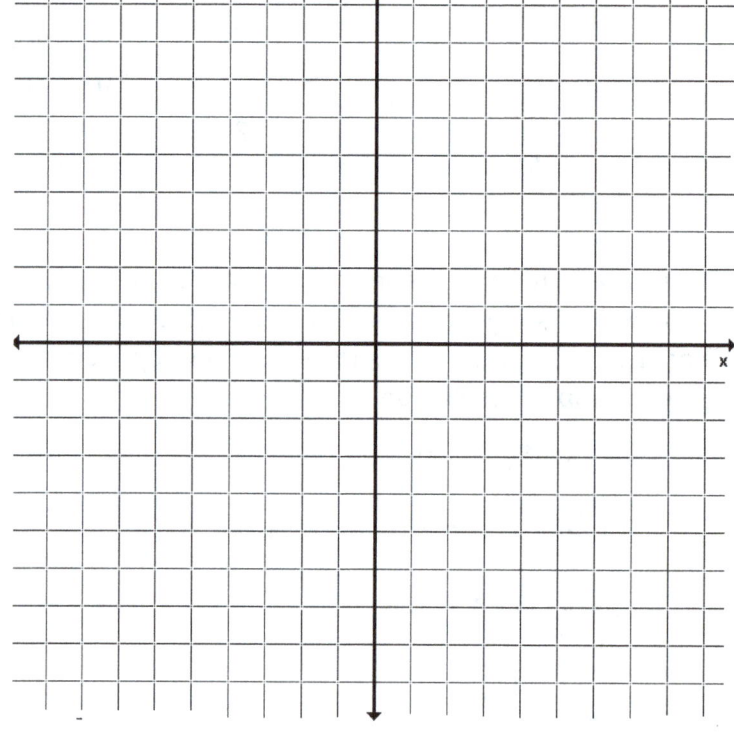

2. Write an algebraic expression for the distance between the following pairs of points in the Cartesian Coordinate Plane: (x, y) and (a, b)

3. Write an equation for all points (x, y) which are a distance 4 units away from the point $(^-3, ^-5)$.

Read and Study 15: The Distance Formula

If I am given a formula, and I am ignorant of its meaning, it cannot teach me anything. But if I already know it, what does the formula teach me?

St. Augustine

In the Class Activity, we essentially asked you to derive a formula for the distance between any two points in the plane. This formula appears often in math texts and is usually called "The Distance Formula". But to us, this is a good example of what St. Augustine was talking about. Knowing a formula without knowing its meaning is useless. And once you understand the meaning of this formula, you realize that memorizing it is pretty pointless, since all it does is applies the Pythagorean Theorem to a right triangle connecting the two points.

So instead of memorizing formula, we think it's more powerful just think of a right triangle or the coordinate grid and use the Pythagorean Theorem whenever you want to compute or write an expression for the distance between two points. And if you end up writing down expressions for distances often enough, eventually you will just write down "the distance formula" as being obvious.

On the other hand, the ability to derive a general formula is a hugely important algebraic skill. One of the key Algebraic Habits of Mind identified by Mark Driscoll is "Abstracting from Computation" which means reasoning about computations independently from the numbers used. Deriving "the distance formula" by thinking about computing the distances on a grid between arbitrary points (a, b) and (x, y) is a great problem to exercise in this kind of abstract thinking.

To help visualize our reasoning, and use what we know and understand about finding distances on a grid, we will draw our two arbitrary points (a, b) and (x, y) somewhere on a grid. We keep in mind that they could be *anywhere*, but if we are going to draw them, we need to draw them *somewhere*. Here's two possibilities. We've put numbers one the axes, but we won't use these numbers to specify the locations of the points, since we want to keep it general.

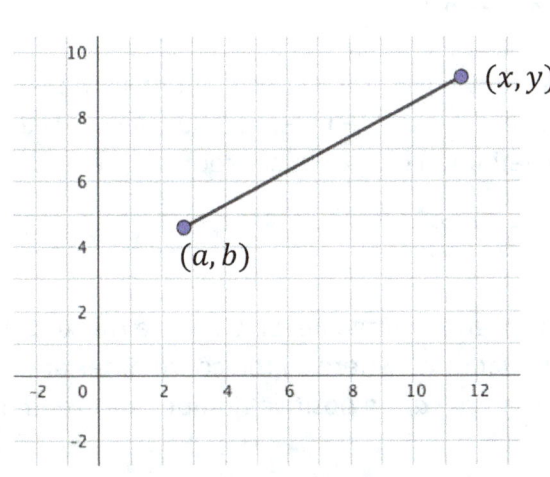 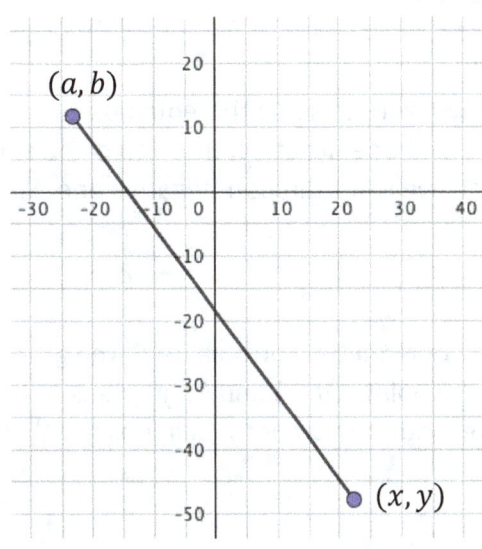

We'll pick the first graph to do our reasoning. To use the Pythagorean Theorem, we'll draw in a right triangle along the horizontal and vertical lines. There's actually two right triangles we could draw (we'll have you consider the other as an exercise), but shown on the right is one possibility. It takes some thought to figure out the coordinates of that other vertex of the triangle. The x-coordinate is the same as the vertex above it, while the y-coordinate is the same as the vertex to its left. Make sure you see why the vertex at the right angle has coordinates (x, b).

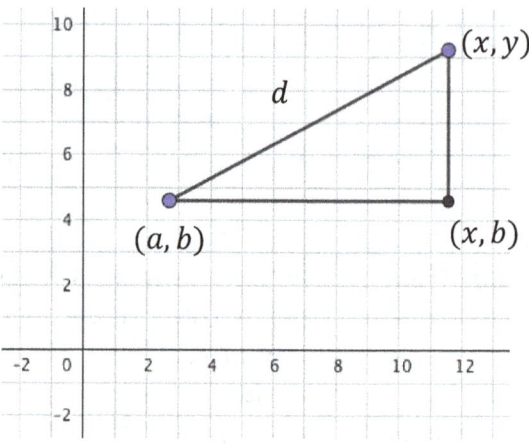

We will let d be the length of the hypotenuse of this triangle. That's the distance between the points that we are trying to find out. Then we can figure out the lengths of the two legs of the triangle. Since these legs are horizontal and vertical, it's easy to find their lengths by reasoning along the grid. The horizontal leg will have length $x - a$, while the vertical leg has length $y - b$.

Don't just take our word for it. Convince yourself we are right! To help with your reasoning, you could imagine specific numbers for x, y, a and b. For example, if x was actually 12 and a was actually 3, then the length of that bottom leg would be the distance between 3 and 12, which you can find by subtracting $12 - 3$. Recognizing that the length of this leg would be the same calculation $(x - a)$ regardless of what numbers x and a were is what is meant by "Abstracting from Computation".

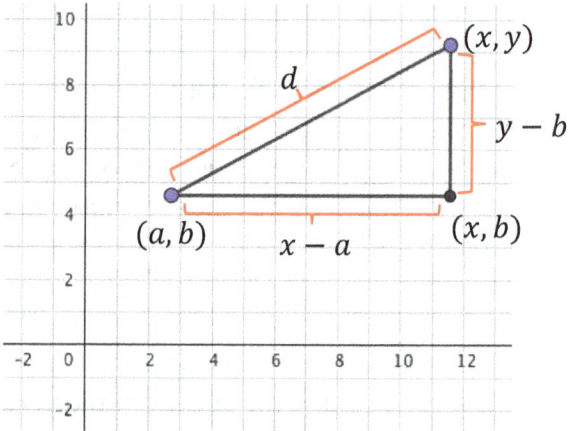

Then by the Pythagorean Theorem, the square of the length of the hypotenuse is equal to the sum of the squares of the two legs. So d^2 will equal the sum of $(x - a)^2$ and $(y - b)^2$. So we get the equation

$$d^2 = (x - a)^2 + (y - b)^2,$$

Now if we were to solve this equation for d, there would in general be two solutions, one positive and one negative, that you could square to get $(x - a)^2 + (y - b)^2$. But since d is a distance, we want the non-negative one. So by the definition of the square root,

$$d = \sqrt{(x - a)^2 + (y - b)^2}.$$

Now we derived this formula using the graph on this page as our guide. So we were thinking about the points (a, b) and (x, y) being in the "first quadrant" where all the coordinates are positive, and the number x being bigger than a, so that $x - a$ is a positive number and $y - b$ is a positive number. But what if the points (a, b) and (x, y) were somewhere else, or if we chose to label them differently, would this affect the formula?

First let's think about what if we had the numbers a and b being bigger than the numbers x and y. Then our diagram would look like the one shown to the right, with the roles of a and x switched, and the roles of x and y swiched. Then the Pythagorean Theorem would result in the equation:

$$d = \sqrt{(a-x)^2 + (b-y)^2}.$$

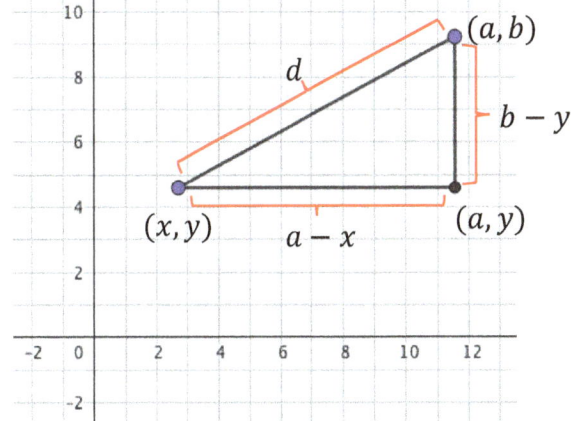

Compare that with our first formula:

$$d = \sqrt{(x-a)^2 + (y-b)^2}.$$

They look different. The first formula had an $(x-a)^2$, and in the second one that was replaced by $(a-x)^2$. Does that make a difference? Well certainly $x - a$ and $a - x$ are different numbers. Thinking about a specific example, $8 - 3$ is a different number than $3 - 8$. But they key here is that when we square them, we will get the same number, since , 5^2 and $(^-5)^2$ are both 25.

We can prove that $(x - a)^2 = (a - x)^2$ in general by using the definitions and laws of algebra. Here's one way:

$$(x - a)^2$$
$$= (x - a) \cdot (x - a) \qquad \text{Definition of exponent}$$
$$= x^2 - 2ax + a^2 \qquad \text{Distributive Law}$$
$$= a^2 - 2xa + x^2 \qquad \text{Commutativity of Addition and Commutativity of Multiplication}$$
$$= (a - x) \cdot (a - x) \qquad \text{Distributive Law}$$
$$= (a - x)^2 \qquad \text{Definition of Exponent}$$

Notice here we did a lot of doing, then undoing. Here's another way, that focuses on the fact that $(x - a)$ and $(a - x)$ have the same absolute value but with opposite signs.

$$(x-a)^2$$

$$= [(^-1) \cdot (^-x + a)]^2 \qquad \text{Distributive Law (Factor out a } ^-1)$$

$$= (^-1)^2 \cdot (^-x + a)^2 \qquad \text{Property of Exponents}$$

$$= (^-x + a)^2 \qquad \text{Since } (^-1)^2 = 1$$

$$= (a - x)^2 \qquad \text{Commutativity of Addition}$$

So the two formulas we derived are equivalent. In practice, this means that if you are using this distance formula, it doesn't matter which point you call (x, y) and which you call (a, b).

There's one more thing we need to think about though before we accept that this formula does indeed generalize all possible cases. So far in our diagrams we have been thinking only about cases where the points had all positive coordinates. Would anything change if some of the coordinates were negative? In the exercises we will ask you to think about this.

Homework Set 15

1) Consider deriving the distance formula using the first diagram in the Read and Study. What instead we had drawn the following right triangle connecting our two points?

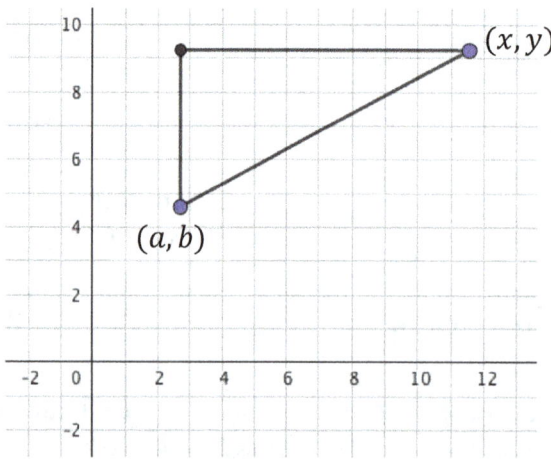

a) What are the coordinates of the third vertex of this triangle?
b) What are the lengths of each leg? How do these lengths compare with the triangle we used in the Read and Study?

2) Consider deriving the distance formula using the second diagram in the Read and Study. Shown are two ways a right triangle could be drawn connecting the points.

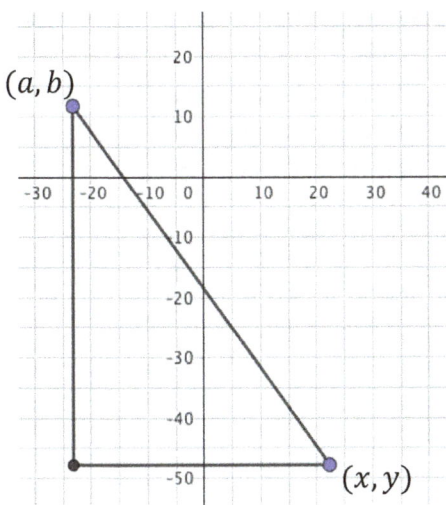
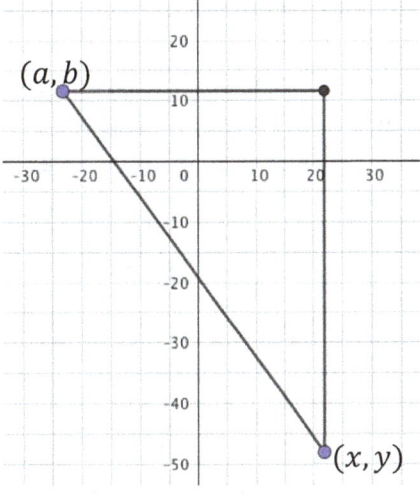

a) Find the coordinates of the third vertex for each triangle.
b) Find the lengths of each leg for each triangle.
c) Use the Pythagorean Theorem to write an equation that gives the length of the hypotenuse. Is this equivalent to the distance formula we derived in the Read and Study? Explain.

3) Find all the points on the line $x = {}^-2$ which are 5 units away from the point $({}^-6, 4\,)$. Do this problem two ways:
   a) Draw the line and the point on grid paper, and find the points by reasoning on the grid.
   b) Note that any point on the line $x = {}^-2$ is of the form $({}^-2, y)$. So write an equation that says the distance between the points $({}^-6, 4\,)$ and $({}^-2, y)$ is 5, and solve that equation for y.

4) Find all of the points on the line $y = x$ that are 7 units away from the point $(4, {}^-3\,)$. Do this problem two ways:
   a) Draw the line and the point on grid paper, and find the points by reasoning on the grid.
   b) Note that any point on the line $y = x$ is of the form (x, x). So write an equation that says the distance between the points $(4, {}^-3\,)$ and (x, x) is 7, and solve that equation for x.

5) Find all of the points on the line $y = 3$ that are 10 units away from the point $({}^-4, 5)$. Do this problem two ways.
   a) Draw the line and the point on grid paper, and find the points by reasoning on the grid.
   b) Note that any point on the line $y = 3$ is of the form $(x, 3)$. So write an equation that says the distance between the points $({}^-4, 5)$ and $(x, 3)$ is 10, and solve that equation for x.

Class Activity 16: Throw Me A Curve

1) A **circle** is the set of points in the plane that are equidistant from a given point in the plane.

   Draw the circle with center $(2, {}^-3)$. and the radius 6. Then write an *equation* that must be satisfied by all the points (x, y) that are a distance 6 away from the point $(2, {}^-3)$. Use your equation to find the precise location of at least eight points on this circle.

   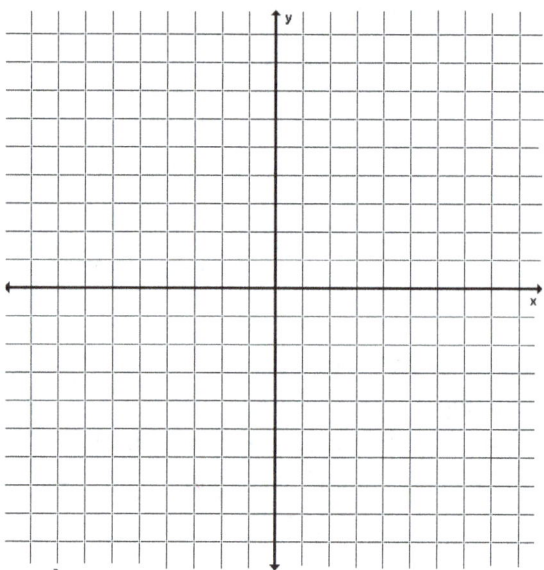

2. A **parabola** is the set of all points in the plane that are equidistant from a given point (the focus) and a given line (the directrix). Use this definition to sketch the parabola with focus $(3, 6)$ and directrix $y = 2$. Write an equation that must be satisfied by any point (x, y) where the distance from (x, y) to the point $(3,6)$ is equal to the distance from the point (x, y) to the line $y = 2$. Use your equation to find the exact location of at least 6 points on the parabola.

   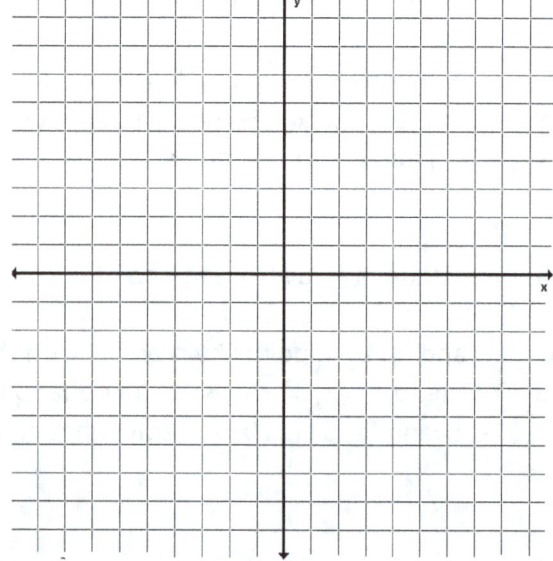

Read and Study 16: Finding Equations for Graphs

A good stock of examples, as large as possible, is indispensable for a thorough understanding of any concept, and when I want to learn something new, I make it my first job to build one.

Paul. R. Halmos

As we defined in an earlier section the **graph** of an equation is the set of points that make the equation true. We already have had a lot of practice of cases where we are given an equation and want to sketch its graph. In other words, to go from an equation to a set of points. Now in this section we are "undoing" that process. We start with a set of points, and try to write down an equation that corresponds to that graph. In the class activity, we had descriptions of very special kinds of geometric graphs (sets of points) that were defined in terms of distances. Lines, parabolas, circles, and ellipses can all be defined in terms of distances in the plane, and these definitions can then be used to write equations that must be satisfied by the graph.

A **circle** is the set of all points that are equidistant from a given point, called the center. The **equation for a circle** is an equation whose solutions are all the points on the circle. In other words, if the ordered pair (x, y) makes the equation true, then (x, y) will be on the circle, and vice versa.

So to derive the equation for a circle, we start by defining the variables x and y so that the point (x, y) is on the circle. Then we will use the definition of the circle to write down what must by true of x and y so that (x, y) is on the circle, namely that the point (x, y) must be that given distance away from the center. This is the same reasoning we will use to derive the equation for an ellipse, or a parabola, or really when solving any problem by writing an equation: define a variable or variables for the quantities that are unknown or can change, and then write down what we know must be true about them.

In the class activity, you wrote an equation for a particular circle, centered at $(2, \bar{\ }3)$ with a radius of 5. But what if we had wanted the equation for a circle centered at $(\bar{\ }5, 7)$ with a radius of 8 ? Or the equation for a circle centered at $(120, 15.6)$ with a radius of 52 ? We could of course do it the same way each time. Or we could just write the equation for all circles at once! By using the same reasoning you used for a particular circle, we can easily write a general equation for the circle centered at the point (h, k), and has radius r. In this context, the numbers h, k, and r are parameters. They are fixed constant numbers, it's just that we haven't specified yet which numbers they are. Then later, after we have our equation written for the general circle, we can substitute any numbers we want for them whenever we want the equation for some particular circle.

In what follows we will show you an example of how you can write down a well-written argument that derives the equation for a circle. (By "derive", we mean to explain where it comes from by using logic to show how the equation results from the definitions involved.)

Suppose we have a circle with center (h, k), and a radius r. Let (x, y) be any point on the circle. Then by the definition of a circle, the distance between the point (x, y) on the circle and the center (h, k) must be r. Using the distance formula we derived in the previous section, that means

$$\sqrt{(x - h)^2 + (y - k)^2} = r.$$

Then by the definition of the square root, we have

$$(x - h)^2 + (y - k)^2 = r^2.$$

And that's it. The equation above is what is known as the standard form for the equation of a circle, with center (h, k), and radius r.

Actually, the first equation we wrote above, with the square root, is also an equation for the circle, so we could have stopped there. If (x, y) is a point on the circle, then x and y will make that equation true. And if x and y make that equation true, then (x, y) will be on the circle. But that equation for a circle is not the usual one you'll find written down. It's more common to simplify this equation to the second form without the square root.

By the definition of the square root the equation $\sqrt{(x - h)^2 + (y - k)^2} = r$ means that r is the *non-negative* number you square to get $(x - h)^2 + (y - k)^2$. But in this context of a circle, the parameter r for the radius of the circle will indeed be a non-negative number, so we can omit stating the provision that $r \geq 0$ and simply write the equation for the circle is $(x - h)^2 + (y - k)^2 = r^2$.

In the class activity, you also found the equation for a graph that is a parabola. Parabolas are likely graphs you are familiar with as being the graph of a quadratic function, but you probably never have considered what it means for a graph to be a parabola.

A **parabola** is the set of all points in the plane that are equidistant from a given point (the focus) and a given line (the directrix). The definition of a parabola is similar to that of a circle in that it's defined in terms of distances, so to write an equation for a parabola, we can apply the same sort of reasoning. We can let (x, y) be any point on the parabola. Then by the definition of a parabola the distance between the point (x, y) on the parabola and the focus must be the same as the distance between the point (x, y) and the directrix. So we can just write an expression for the distance between (x, y) and the focus, and set that equal to an expression for the distance between (x, y) and the directrix. We won't write down a general formula, since we don't want you to simply plug values into a formula. Instead we want you to go through the reasoning involved in writing an equation by yourself, since it's that reasoning that's important for you to learn.

Homework Set 16

1) A circle has a radius of 5 and is centered at $(^-2, 4)$. Find the exact locations of 12 points on this circle. (Note: you don't need to solve equations to do this. You can use reasoning on grid paper and symmetry to find this many points. Remember 3-4-5 right triangles!)

2) Your task is to derive an equation for the given curves, starting from the definition. Start by making an accurate sketch of the curve on graph paper. Then by using the definition of the curve and the concept of distance write down an equation for any point (x, y) on the curve. Then with the help of your equation, find and label at least 8 points on each curve.

   a) The circle with a diameter with endpoints $(^-1, 4)$ and $(7, 8)$.

   b) The circle centered at $(5, ^-3)$ that goes through the point $(^-2, 1)$.

   c) The parabola with directrix $y = ^-6$ and focus at $(4, 0)$

   d) The parabola with a vertex at $(0, 6)$ and a focus at $(0, 2)$. (Suggestion: First figure out where the directrix must be.)

3) Plot some points that are equidistant between the points $(^-6, 2)$ and $(4, 8)$. Try to find at least two. Then write an equation that must be satisfied by all the points (x, y) that are equidistant between the points $(^-6, 2)$ and $(4, 8)$, and use your equation to find more points and sketch the graph.

4) Find the directrix and focus for parabola defined by the equation $y = x^2$. Hint: graph the equation $y = x^2$. Since the parabola has a vertex at $(0, 0)$, we know that the focus should be at $(0, b)$ and the directrix should be at $y = -b$, for some number b. Use the fact that $(1, 1)$ is on the parabola to determine the value for b.

Class Activity 17: Squashed Circles

An **ellipse** is the set of all points in the plane such that the *sum* of the distances from two given points (the foci – that's plural for focus) is constant.

Sketch the ellipse with foci $(^-3, 0)$ and $(3, 0)$ and a distance sum of 10 units. Then write down an equation that must be true for any point (x, y) that satisfies the definition of this ellipse. Use your equation to find the precise location of at least eight points on the ellipse.

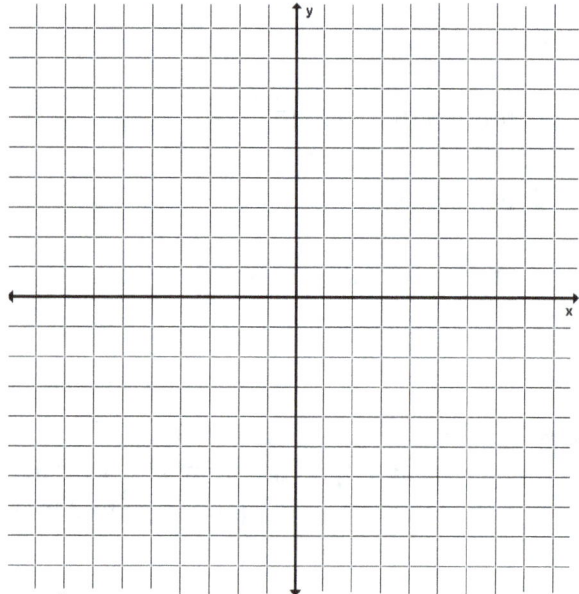

Read and Study 17: Finding Equations for Ellipses

Throw a pumpkin in the air and it will come down squash.

An **ellipse** is the set of all points in the plane such that the *sum* of the distances from two given points (the foci – that's plural for focus) is constant. Notice the similarity between the definition of a circle and the definition of an ellipse. An ellipse is just like a circle, but with two foci, instead of just one center. In fact, we like to think of a circle as being just a special case of an ellipse where the two foci are the same point.

To derive the equation for an ellipse, we can use the same reasoning as we did for the circle. However, instead of using parameters for the location of the foci and the total distance, let's consider another specific example of an ellipse, this time with foci at (3,4) and (3,10), and a sum of the distances 12.

We'll start by sketching the features we know on a graph, and then use the definition to write an equation. Then we can use that equation to find more points on the ellipse and make a more accurate graph. On the right is accurate sketch of the ellipse we made using GeoGebra. (If your instructor hasn't told you yet about GeoGebra, go to www.geogebra.org now. It's really great software. Easy to use, and free!)
We'll draw the ellipse in blue. (We've marked the foci in red, since the ellipse). Now when sketching this by hand, we won't yet know ellipse goes through. We can figure out that (3,1) a point on the ellipse, since (3,1) is a distance of 3 away from one focus, and a distance 9 away from the other focus, the total distance is 12. Similarly, we can argue that the point (3,13) at the top is on the ellipse.

Looking at the graph above, it looks like perhaps the point (6,2) might be on the ellipse. As an exercise to better understand the definition of an ellipse, let's see if this is the case. The point (6,2) is a distance of $\sqrt{13}$ away from (3,4), and a distance of $\sqrt{73}$ away from (3,10). So the sum of the distances away from the foci is $\sqrt{13} + \sqrt{73} \approx 12.15$. (Verify these calculations!). So close, but no cigar. In order to be *on* the ellipse, the sum of the distances must be exactly 12. Since the distance was a little more than 12, we can see that (6,2) is a little bit outside the ellipse, and not right on it.

To derive an equation for this ellipse, we can essentially do the same calculations as we did in the previous paragraph, but using an arbitrary point that we define to be on the ellipse. Let (x, y) be a point on the ellipse with foci at (3,4) and (3,10), and a sum of the distances 12. To help us visualize the argument, we will imagine the point (x, y) at some arbitrary point on the ellipse. We realize that (x, y) can be *anywhere* on the ellipse, but to draw a picture, we need to put it *somewhere*. Here's just two of an infinite number of possibilities.

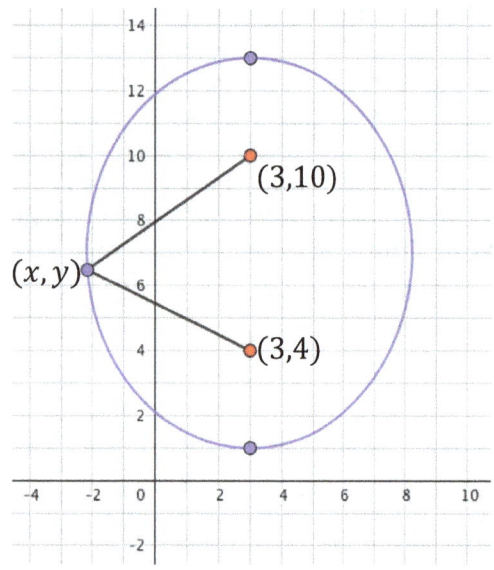 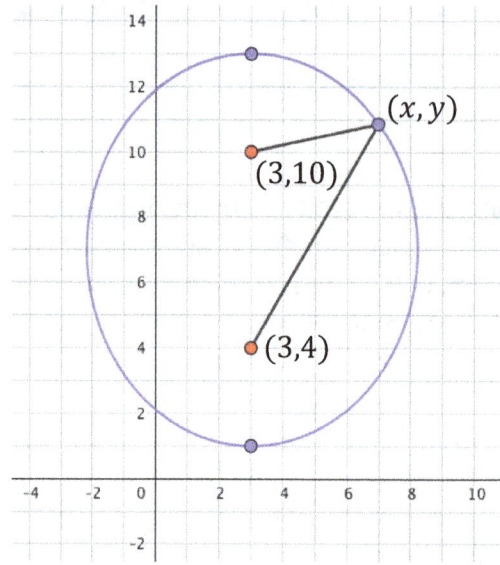

We've also drawn in line segments showing the distance from the point (x, y) to each focus. By using the distance formula, the distance from (x, y) to $(3,4)$ is $\sqrt{(x-3)^2 + (y-4)^2}$, and the distance from (x, y) to $(3,10)$ is $\sqrt{(x-3)^2 + (y-10)^2}$. By the definition of the ellipse, the sum of these two distances must be 12. So if (x, y) is on the ellipse, we must have

$$\sqrt{(x-3)^2 + (y-4)^2} + \sqrt{(x-3)^2 + (y-10)^2} = 12.$$

So there we have an equation for this ellipse. We've shown that if (x, y) is any point on the ellipse, it must satisfy this equation. In general, if we wanted to find the equation for an ellipse with foci (p_1, q_1) and (p_2, q_2) and constant sum d of the distances to the foci, we could make the same argument and would end up with the equation

$$\sqrt{(x-p_1)^2 + (y-q_1)^2} + \sqrt{(x-p_2)^2 + (y-q_2)^2} = d.$$

This is a general equation for an ellipse with foci (p_1, q_1) and (p_2, q_2). Make sure you understand this formula. We wrote it here so we can talk about it, not so that you would memorize it. Remember what St. Augustine said. Knowing this formula without understanding it is useless. And if you understand the formula, you don't need to memorize it.

So what's the value of having an equation for the ellipse? Mainly so we can use it to find exact locations of points on it. (How do you think the computer was programmed to draw that nice picture of ours? Essentially, it used the equation!).

We'll use the equation we derived to find the exact location of some more points on our ellipse. Let's find the left and rightmost extreme points. We know from symmetry that this occurs when $y = 7$. But what are the x-values there? We can substitute 7 for y into our equation and solve for x. Then we'd get

$\sqrt{(x-3)^2 + (y-4)^2} + \sqrt{(x-3)^2 + (y-10)^2} = 12$		Equation for our ellipse
$\sqrt{(x-3)^2 + (7-4)^2} + \sqrt{(x-3)^2 + (7-10)^2} = 12$		Substitute $y = 7$
$\sqrt{(x-3)^2 + 9} + \sqrt{(x-3)^2 + 9} = 12$		$(7-4)^2 = 9$ and $(7-10)^2 = 9$
$2 \cdot \sqrt{(x-3)^2 + 9} = 12$		Distributive Law (Combining Like Terms)
$\sqrt{(x-3)^2 + 9} = 6$		Property of Equality: multiply by $\frac{1}{2}$
$(x-3)^2 + 9 = 36$		Definition of the Square Root
$(x-3)^2 = 27$		Property of Equality: subtract 9
$x - 3 = \sqrt{27}$ or $x - 3 = {}^-\sqrt{27}$		The numbers whose square is 27
$x = 3 + \sqrt{27}$ or $x = 3 - \sqrt{27}$		Property of Equality: add 3

So when $y = 7$, we get two solutions $x = 3 + \sqrt{27}$, which is approximately 8.196, and $x = 3 - \sqrt{27}$, which is approximately $^-$2.196. We've plotted the points $(3 + \sqrt{27}, 7)$ and $(3 - \sqrt{27}, 7)$ on the graph below.

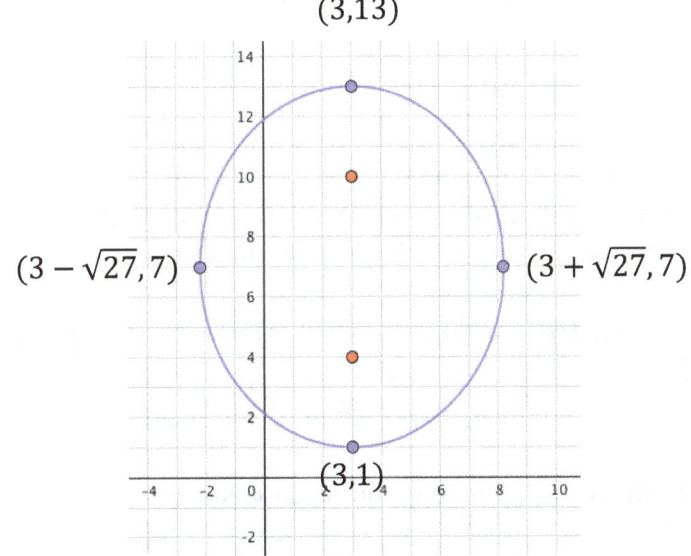

Let's now find the points that are across from the foci. Since the foci are at a y value of 4, we'll substitute $y = 4$ into our equation for the ellipse.

$\sqrt{(x-3)^2 + (y-4)^2} + \sqrt{(x-3)^2 + (y-10)^2} = 12$		Equation for our ellipse
$\sqrt{(x-3)^2 + (4-4)^2} + \sqrt{(x-3)^2 + (4-10)^2} = 12$		Substitute $y = 4$

$$\sqrt{(x-3)^2} + \sqrt{(x-3)^2 + 36} = 12 \qquad \text{Since } (4-4)^2 = 0 \\ \text{and } (4-10)^2 = 36$$

We pause now for an important discussion. We know it is *very* tempting for many students to want to try to undo those square roots by "squaring everything" and write

$$(x-3)^2 + (x-3)^2 + 36 = 144,$$

but this equation is **not** equivalent to ours (hence the red "warning" ink), for several important reasons.

First, remember that **there is no such property of equality that says you can square both sides of an equation**. To deal with square roots, we need to use the definition of the square root. Second, the left side of the equation is the sum of two terms. And squaring the sum of two terms is not equivalent to adding the squares of each term. Remember that **exponents do not distribute**. For example:

$$(1+2)^2 \neq 1 + 4, \text{ and}$$
$$(\sqrt{25} + \sqrt{9})^2 \neq 25 + 9.$$

Check these out yourself by computing each side.

In general,

$$(a+b)^2 \neq a^2 + b^2,$$
$$\text{and}$$
$$(\sqrt{a} + \sqrt{b})^2 \neq a + b.$$

If we want to simplify squaring an expression with two terms, we need to use the distributive law. That's because squaring is multiplication, and terms are added, and it's the distributive law that relates multiplying across addition. So, using the distributive law, we could write

$$(a+b)^2 = (a+b) \cdot (a+b) = a^2 + 2ab + b^2,$$
$$\text{and}$$
$$(\sqrt{a} + \sqrt{b})^2 = (\sqrt{a} + \sqrt{b}) \cdot (\sqrt{a} + \sqrt{b}) = a + 2\sqrt{a}\sqrt{b} + b$$

Check those out yourself now by using the distributive law. Really, do it!

So back to solving $\sqrt{(x-3)^2} + \sqrt{(x-3)^2 + 36} = 12$ for x. There is something we can do to simplify this equation before we proceed, namely, to rewrite the expression $\sqrt{(x-3)^2}$. This expression means that we take a number x, subtract 3, square the result, then take the square root. But doesn't taking the square root just undo the squaring? Well, yes, but assuming $x - 3$ is positive, and here's why:

By the definition of the square root, $\sqrt{(x-3)^2}$ is the non-negative number you square to get $(x-3)^2$. So what number(s) do you square to get $(x-3)^2$? Well obviously $(x-3)$ works. But so does $^-(x-3)$. (Check it out!). So which one is it? By the definition, the square root is the non-negative one. It is tempting to say that $(x-3)$ is the positive one and $^-(x-3)$ is the negative one, but that will depend on what number x is. If x is the number 2, then $(x-3)$ is negative and $^-(x-3)$ is positive. But if x is bigger than 3, then $(x-3)$ is the positive one. Since we don't yet know what number x is, let's assume, for now, that $\sqrt{(x-3)^2}$ is $(x-3)$.

$$\sqrt{(x-3)^2} + \sqrt{(x-3)^2+36} = 12$$

$$x-3 + \sqrt{(x-3)^2+36} = 12 \qquad \text{Assuming } \sqrt{(x-3)^2} \text{ is } (x-3).$$

$$\sqrt{(x-3)^2+36} = 15-x \qquad \text{Add 3 and subtract x to both sides}$$

$$(x-3)^2+36 = (15-x)^2 \qquad \text{Definition of the square root, assuming } x \leq 15$$

$$x^2-6x+9+36 = 225-30x+x^2 \qquad \text{Distributive Law}$$

$$-6x+45 = 225-30x \qquad \text{Subtract } x^2 \text{ from both sides}$$

$$24x = 180 \qquad \text{Add } 30x-45 \text{ to both sides}$$

$$x = \frac{180}{24} = \frac{15}{2} \qquad \text{Divide by 24}$$

So that's pretty cool. Exactly 7.5. But wait, shouldn't there be another solution? Looking at the graph of our ellipse, there are two points on the ellipse where $y = 4$. We substituted $y = 4$ into the equation, but only found the one x value. What happened? Well, remember when we rewrote $\sqrt{(x-3)^2}$ as $(x-3)$? That assumed $(x-3)$ was positive, so x was greater than 3 and that's what we found, the point on the ellipse where y is 4 and x is greater than 3. The other point we are looking for is where x is less than 3. In that case, $\sqrt{(x-3)^2}$ will be $^-(x-3)$, since $^-(x-3)$ will then be positive. Then the calculations would look like this:

$$\sqrt{(x-3)^2} + \sqrt{(x-3)^2+36} = 12$$

$$^-(x-3) + \sqrt{(x-3)^2+36} = 12 \qquad \text{Assuming } \sqrt{(x-3)^2} \text{ is } ^-(x-3).$$

$$\sqrt{(x-3)^2+36} = x+9 \qquad \text{Add } x-3 \text{ to both sides}$$

$$(x-3)^2+36 = (x+9)^2 \qquad \text{Definition of the square root, assuming } x \geq {}^-9$$

$$x^2 - 6x + 9 + 36 = x^2 + 18x + 81 \qquad \text{Distributive Law}$$

$$-6x + 45 = 18x + 81 \qquad \text{Subtract } x^2 \text{ from both sides}$$

$$^-36 = 24x \qquad \text{Add } 6x - 81 \text{ to both sides}$$

$$\frac{^-3}{2} = \frac{^-36}{24} = x \qquad \text{Divide by 24}$$

So the other point on the ellipse when y is 4 is when x is exactly $^-1.5$. We'll add those points to our graph.

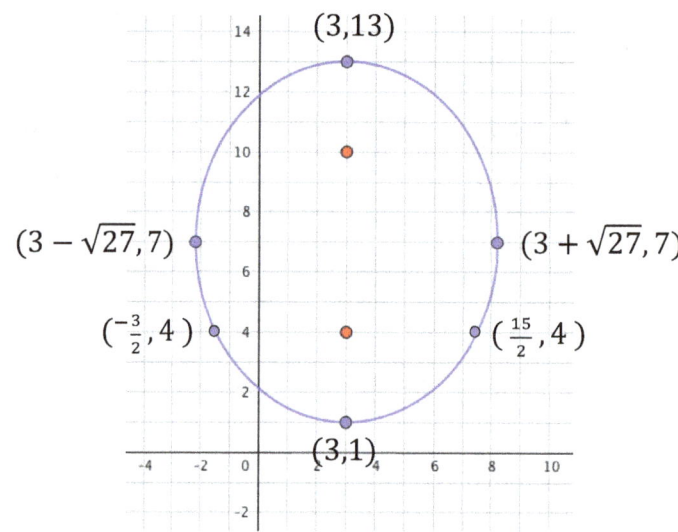

Now, by symmetry, we can predict that the points across from the other focus at $y = 10$ will also be at $x = \frac{15}{2}$ and at $x = \frac{^-3}{2}$. And this is easily confirmed by the equation. If you substitute $y = 10$ into the equation, you will get the same equation we had before, and thus will get the same two solutions.

Now, we've been working this this ellipse for a while now, but let's hit home an important point we made earlier about solving equations with square roots by considering finding some other points on this ellipse. The truth is that the points we have found so far are the only "easy" ones to find. We could find more, but solving the equation becomes a bit more difficult. For example, suppose we wanted to find the x values when $y = 2$.

$$\sqrt{(x-3)^2 + (y-4)^2} + \sqrt{(x-3)^2 + (y-10)^2} = 12 \qquad \text{Equation for our ellipse}$$

$$\sqrt{(x-3)^2 + (2-4)^2} + \sqrt{(x-3)^2 + (2-10)^2} = 12 \qquad \text{Substitute } y = 2$$

$$\sqrt{(x-3)^2 + 4} + \sqrt{(x-3)^2 + 64} = 12 \qquad \begin{array}{l}\text{Since } (2-4)^2 = 4 \\ \text{and } (2-10)^2 = 64\end{array}$$

Have a look at that last equation. What would you do next to try to solve it? As we explored earlier, we can not simply this equation by just squaring each term. This equation is NOT equivalent to

$$(x-3)^2 + 4 \;+\; (x-3)^2 + 64 \;=\; 144.$$

Nor can we take the square root of each term under the square roots. We know this is *very* tempting, since $(x-3)^2$, 4 and 64 are all squares, but the last equation is NOT equivalent to

$$(x-3) + 2 \;+\; (x-3) + 8 = 12$$

for all the same reasons we talked about a few pages ago. **Square roots do not distribute!**

To undo those square roots, we would need to use the definition of the square root, and we would need to apply it to one of these square roots at a time. It is tedious, we know, but it's the only way. We will show you what that would look like, not because we expect you to do this, or to memorize this procedure, but simply to show you that it can be done, and that every step is justified by a definition or Law of Algebra.

$\sqrt{(x-3)^2 + 4} \;+\; \sqrt{(x-3)^2 + 64} \;=\; 12$	
$\sqrt{(x-3)^2 + 4} \;=\; 12 - \sqrt{(x-3)^2 + 64}$	Property of Equality: Add $-\sqrt{(x-3)^2 + 64}$
$(x-3)^2 + 4 \;=\; \left(12 - \sqrt{(x-3)^2 + 64}\right)^2$	Def. of square root, if $\sqrt{(x-3)^2 + 64} \leq 12$
$(x-3)^2 + 4 \;=\; \left(12 - \sqrt{(x-3)^2 + 64}\right) \cdot \left(12 - \sqrt{(x-3)^2 + 64}\right)$	Def. of Exponent
$(x-3)^2 + 4 \;=\; 144 - 24 \cdot \sqrt{(x-3)^2 + 64} + (x-3)^2 + 64$	Distributive Law
$(x-3)^2 + 4 \;=\; 208 - 24 \cdot \sqrt{(x-3)^2 + 64} + (x-3)^2$	Combine Like Terms
$0 \;=\; 204 - 24 \cdot \sqrt{(x-3)^2 + 64}$	Prop. of Equality: add $-(x-3)^2 - 4$
$24\sqrt{(x-3)^2 + 64} \;=\; 204$	Prop of Equality: add $24\sqrt{(x-3)^2 + 64}$
$\sqrt{(x-3)^2 + 64} \;=\; \frac{17}{2}$	Property of Equality: multiply by $\frac{1}{24}$
$(x-3)^2 + 64 \;=\; \left(\frac{17}{2}\right)^2$	Def. of Square Root

$$x^2 - 6x + 9 + 64 = \frac{289}{4}$$ Distributive Law, and $17 \cdot 17 = 289$

$$x^2 - 6x + \frac{3}{4} = 0$$ Property of Equality

$$x = \frac{6 + \sqrt{(-6)^2 - 4 \cdot \left(\frac{3}{4}\right)}}{2} \text{ or } x = \frac{6 - \sqrt{(-6)^2 - 4 \cdot \left(\frac{3}{4}\right)}}{2}$$ Quadratic Formula

$$x = \frac{6 + \sqrt{33}}{2} \text{ or } y = \frac{6 - \sqrt{33}}{2}$$

$$x \approx 5.87 \text{ or } x \approx 0.1277$$

Go ahead and find and label these points on our graph of the ellipse.
Wow that was long calculation. It turns out that the equation for an ellipse that we derived,
$$\sqrt{(x - p_1)^2 + (y - q_1)^2} + \sqrt{(x - p_2)^2 + (y - q_2)^2} = d$$
with foci (p_1, q_1) and (p_2, q_2) and constant sum d, can be a bit difficult to work with if you have to deal with two different square roots. But it has the advantage that it's easy to derive from the definition of an ellipse, and it works regardless of where the two foci are, even if the ellipse is "tilted" with respect to the x and y axes.

However, we can derive a simpler equation for an ellipse if instead of using the locations of the foci, we use the location of the center and the lengths of the shortest and longest radii.

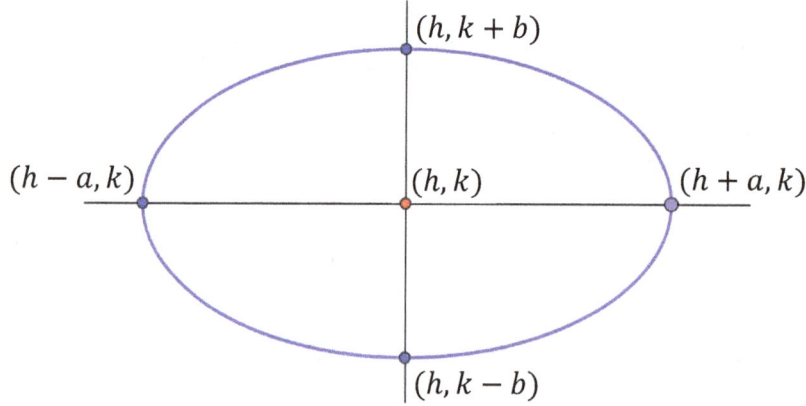

In the diagram above, we have shown an ellipse centered at (h, k), with the longest and shortest radii being a and b. Technically these are called the lengths of the semi-major axis (half the longest diameter) and the semi-minor axis (half the shortest diameter). If the center of the ellipse is (h, k) and a and b are the lengths of the semi-axes, then, assuming these axes are aligned horizontally and vertically, the equation for the ellipse is

$$\frac{(x - h)^2}{a^2} + \frac{(y - k)^2}{b^2} = 1$$

A nice feature of the form $\frac{(x-h)^2}{a^2} + \frac{(y-k)^2}{b^2} = 1$ for the equation of an ellipse is not just that it no longer involves square roots, its also that you can see the connection with the general equation for a circle $(x - h)^2 + (y - k)^2 = r^2$, which we will ask you to consider in the exercises.

To understand this form of the equation, let's figure out what (h, k) and a and b would be for the ellipse in our example. From the symmetry, we can see that the center is at $(3,7)$. The length of the semi-minor axis (in this case in the horizontal x direction) is $a = \sqrt{27}$, while the length of the semi-major axis (in this case in the vertical y direction) is $b = 6$. Hence an equation for this ellipse would be

$$\frac{(x-3)^2}{27} + \frac{(y-7)^2}{36} = 1$$

Recall that the first equation we derived for this ellipse was

$$\sqrt{(x-3)^2 + (y-4)^2} + \sqrt{(x-3)^2 + (y-10)^2} = 12.$$

It would be another rather a messy problem to show that these two equations are equivalent. In fact, it would look a lot like that long calculation a few pages ago. Instead, we will ask you in the exercises to use this new form of the equation to find points on the ellipse we have already located to see that it indeed does generate the same ellipse.

Homework Set 17

1) In the read and study, we claimed that the equations
$$\sqrt{(x-3)^2} + \sqrt{(x-3)^2 + 36} = 12 \quad \text{and}$$
$$(x-3)^2 + (x-3)^2 + 36 = 144$$
are not equivalent. Prove this by finding all solutions to the second equation and showing that these do not give you the points on the ellipse when $y = 4$.

2) Also the read and study, we claimed that the equations
$$\sqrt{(x-3)^2 + 4} + \sqrt{(x-3)^2 + 36} = 12 \quad \text{and}$$
$$(x-3)^2 + 4 + (x-3)^2 + 36 = 144$$
are not equivalent. Prove this by finding all solutions to the second equation and showing that these do not give you the points on the ellipse when $y = 2$.

3) Yet again in the read and study, we claimed that the equations
$$\sqrt{(x-3)^2 + 4} + \sqrt{(x-3)^2 + 36} = 12 \quad \text{and}$$
$$(x-3) + 2 + (x-3) + 6 = 12$$

are not equivalent. Prove this by finding all solutions to the second equation and showing that these do not give you the points on the ellipse when $y = 2$.

4) Consider the second equation for the ellipse in the Read and Study, $\frac{(x-3)^2}{27} + \frac{(y-7)^2}{36} = 1$.
   Use this equation to:
   a) Find the x-values on the ellipse when $y = 4$.
   b) Find the x-values on the ellipse when $y = 2$.
   c) Find the points on the ellipse when $x = 1$.

5) Derive an equation for the ellipse centered at $(6,0)$ with a focus at $(4,0)$ and a vertex at $(0,0)$. Your task is to derive the equation starting from the definition of an ellipse. Do not simply substitute values into a formula from our text. Start by making an accurate sketch of the curve on graph paper. Then by using the definition of the curve and the concept of distance write down an equation for any point (x, y) on the curve. Then with the help of your equation, find and label at least 8 points on each curve.

6) In the read and study we said that the general equation for an ellipse with center (h, k) and length of the semi-major and semi-minor axes a and b is $\frac{(x-h)^2}{a^2} + \frac{(y-k)^2}{b^2} = 1$.
   a) Use this form to write an equation for the ellipse from the Class Activity.
   b) Use this new equation to verify the points on the ellipse you have already found in the Class Activity.
   c) Use this new equation to find the exact location of 4 more points on the ellipse.

7) An ellipse has equation $\frac{x^2}{49} + \frac{y^2}{16} = 1$.
   a) Sketch a graph of this ellipse on grid paper. Find the exact locations of 8 points.
   b) Find d the sum of the distances from any point to the two foci. (Hint: consider how far away the vertex on the x-axis is away from the two foci.)
   c) Figure out the exact location of the two foci. (Hint: consider how far the vertex on the y-axis is away from the two foci.)
   d) Use the definition of the ellipse to write another equation for this ellipse involving the foci and the sum of the distances away from the foci.

8) Under what conditions on a, b and r will the equations
$$\frac{(x-h)^2}{a^2} + \frac{(y-k)^2}{b^2} = 1$$
   and
$$(x-h)^2 + (y-k)^2 = r^2$$

   be identical? What does this say about the relationship between a circle and an ellipse?

Chapter Three

Analyzing Functions

Class Activity 18a: Directory Assistance

A **function** is a relation in which each member of one set (called the **domain**) corresponds to exactly one member of another set (called the **range**).

In this activity, we will explore the concept of a function using the university online directory. First, discuss in your group some possible domains and ranges for the directory. For each domain and range you describe, discuss whether or not you think the directory would represent a function between that domain and range, and why.

Now your group should consider the university directory as a relationship between a **domain of names** and a **range of telephone numbers.**

a. Assume that there are **no unlisted numbers, no multiple phone lines** for one person, and **no roommates or officemates** with the same phone number. Would the directory be a function between the set of names and the set of phone numbers? Why or why not?

b. Suppose again that there are no unlisted numbers, and no multiple phone lines for one person, but that you allow roommates to have the same telephone number. In this case, would the directory be a function between the set of names and the set of phone numbers? Why or why not?

c. Suppose instead you allow unlisted numbers, but no multiple lines or roommates. In this case, would the directory be a function between the set of names and the set of phone numbers? Why or why not?

d. Suppose instead you allow multiple lines, but no roommates or unlisted numbers. In this case, would the directory be a function between the set of names and the set of phone numbers? Why or why not?

Class Activity 18b: Function Machines Revisited

Each function machine below shows a sequence of operations from left to right. For each:
  a. Find the intermediate outputs and the final output if you input the number x.
  b. Find all the real numbers that if they were inputs into this function, they would not give a real number output.
  c. Find all the real numbers that could never be an output of this function.
  d. Specify all the real numbers that are in the domain and all the real numbers that are in the range of this function.

1.

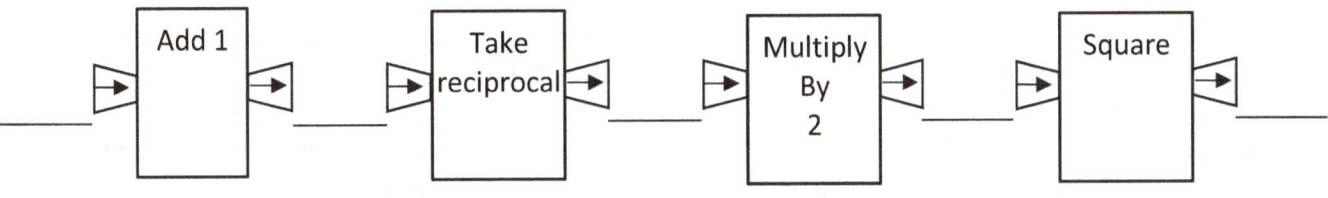

$f(x) =$

Domain: Range:

2.

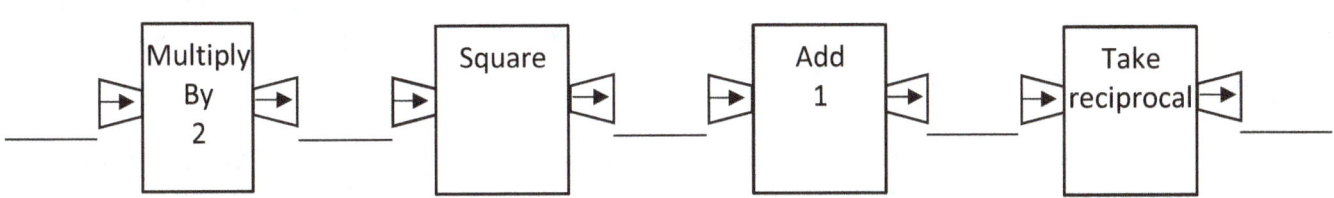

$f(x) =$

Domain: Range:

135

Read and Study 18: Function Definitions

Success is more a function of consistent common sense than it is of genius.

An Wang

The concept of the function is one of the most important concepts in this course. We will start by understanding the definition:

> A **function** is a relationship between two sets in which each member of one set (called the **domain**) corresponds to exactly one member of another set (called the **range**).

In the Handshake problem from the first section of this book, there is a functional relationship between the number of members of the club and the number of handshakes that take place. For each number of members, there is exactly one number of required handshakes. If we let the variable n represent the number of club members, let the variable H represent the number of handshakes required, and let the function f represent the rule that assigns each number of members to a number of handshakes, then we can say that $H = f(n)$.

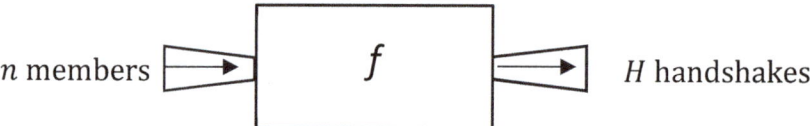

In the Directory Assistance class activity, we explored whether the relation between a domain of names and a range of telephone numbers was a function. In order for us to be able to say the relation with this domain and range is a function, each name in the domain has to correspond to one and only one telephone number in the range. If there is a name that has no number then this relation is not a function. Also, if there is a name that has more than one number, then this relation is not a function.

But if there *were* two or more names that have the same number, then the relation can still be a function between the domain of names and range of numbers. The definition of a function says *only* that each member of the *domain* corresponds to exactly one member of the *range*. It does not say that each member of the range has to correspond to exactly one member of the domain. So a key idea in the function definition is that there is a specified direction to this relationship, and that direction is from the domain to the range.

To help keep this direction from domain to range in mind, a very powerful way to think of a function is the **machine concept.** Think of a function as a machine that accepts inputs from the domain and spits out outputs in the range. Inside the machine is the rule that considers each input and decides which output to produce from that input. We can make a diagram of this concept like we did for the handshake problem above.

With this viewpoint, the **domain** is the set of all possible inputs in to the function, and the **range** is the set of all possible outputs out of the function using elements of the domain as inputs.

In a real-world setting, determining whether your school's student directory is a function between names and numbers can be messy, and depend a lot on assumptions you make. But the point of the exercise was not to be able to get a definitive answer but to better understand the mathematical definition of a function.

A function is a specific type of relationship between two variables. The variable whose values are elements of the domain is called the **independent variable** for the function, and the variable whose values are elements of the range is called the **dependent variable** for the function. *Why does it makes sense to call the domain variable independent and the range variable dependent?* In standard notation, x is used as an independent variable, and y as the dependent variable. Using these variables, a simple machine diagram for a function is as shown.

$$x \longrightarrow \boxed{f} \longrightarrow y$$

The idea represented by this diagram is more commonly written as an equation $y = f(x)$. Here, the symbol f is the name of the function and the notation $f(x)$ means that the variable x is the input into the function called f.

> **Function Notation**
>
> $y = f(x)$ means that y is the output when x is the input into the function called f.

More about Functions. Let's consider another "real world" setting, this time when both variables are numbers. Suppose you are hiking on a hilly hiking trail. Let the variable x be the elapsed time of your hike, and let the variable y be the elevation above sea level. In this situation, would your elevation be a function of the elapsed time? In order to answer this, we need to ask whether each elapsed time would correspond to one and only one elevation. This seems right: at any particular time, you would be at one particular elevation.

It's quite common that a relation that is a function in one direction is not a function in the opposite direction. For example, in the hilly hiking situation, if we instead considered the inputs to be the elevation and the outputs to be the times, then we'd say the relation would not be a function. Since the trail was hilly, there would be elevations that correspond to two different times (one when doing up the hill, and another when going down). So there would be an input (an elevation) that corresponded to more than one output (time), contradicting the definition of a function. So you see that the choice of which variable is being considered the input and which the output is crucial. In our hilly hiking example, the elevations are a function of the times, but the times are not a function of the elevations.

In cases where the domain and range are sets of real numbers then we can make a graph of the function. **The standard convention when making the graph of a function is to put the independent (domain) variable on the horizontal axis, and the dependent (range) variable on the vertical axis.** Take a look at a graph of the following function:

$y = {}^-11x^2 + 66x, \quad 0 \leq x \leq 6.$

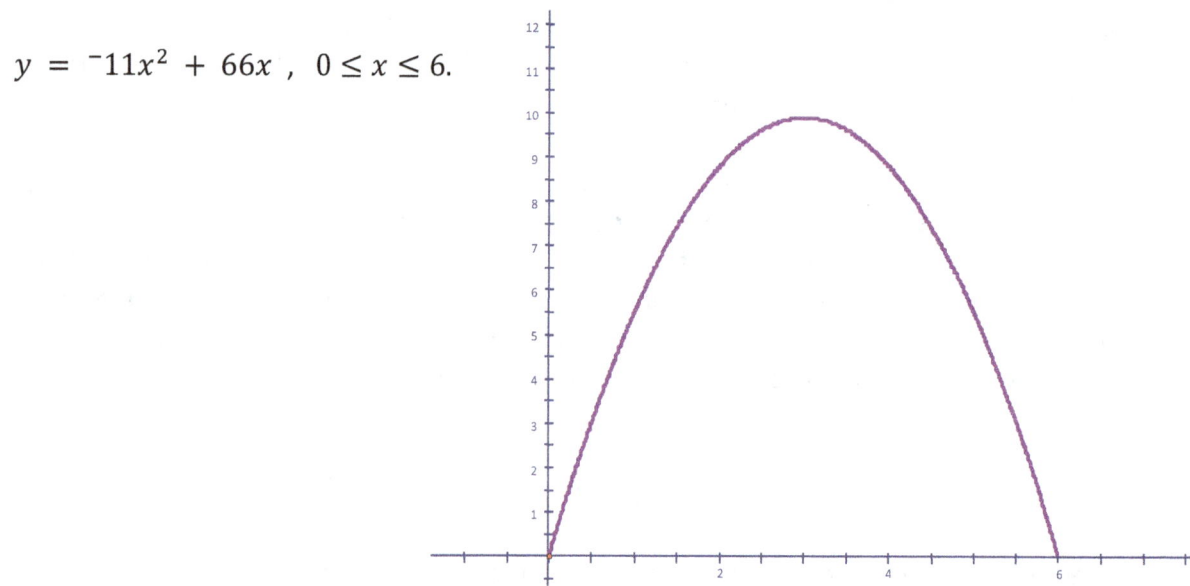

Now, *study how the graph illustrates the relationship*. For example, the point above $x = 2$, has a height or y value of 88. Note that we can tell it's a functional relationship because at each x-value there is only one y-value. Notice how we have specified a domain for this function. We said the relationship is valid for x-values from 0 to 6. If a domain is not explicitly stated, then we usually assume the biggest domain that makes sense for the situation.

The graph can be helpful in determining the range. Recall the range of a function is the set of all possible outputs using elements of the domain as inputs. We can see that the smallest value y takes is zero, which happens when x is either 0 or 6. The largest value for y looks like it occurs when x is 3. Letting the input be 3, we calculate an output of ${}^-11 \cdot (3)^2 + 66 \cdot 3$, which is 99. Furthermore, we can see by the continuity of the graph that the output y could be any real number between 0 and 99. So when the domain of this function is $0 \leq x \leq 6$, the range is $0 \leq y \leq 99$.

We've seen that functions between numerical variables can be represented in several ways: in words, by numeric tables, with an algebraic formula, and with a graph. Let's see all these representations for a specific example.

First, the words: Suppose that I'm earning 12% annual interest in my checking account and I put $100 in there for eight years. Let's consider the amount of money in my account as a function of time.

Here is that data as a numeric table:

Year	Amount
0	$ 100.00
1	$ 112.00
2	$ 125.44
3	$ 140.49
4	$ 157.35
5	$ 176.23
6	$ 197.38
7	$ 212.07
8	$ 247.60

Do the calculations to check that this data is correct.

Here is the amount of money in my account captured by an algebraic formula:

$$y = 100 \cdot (1.12)^x$$

Check that these are all the same function and check that they give the above table values. Do you see where that 1.12 comes from?

Finally, here is a graph showing the relationship between time (on the x-axis) and the money in my account on the y-axis. (Think of a graph as a picture of the numeric or algebraic models.)

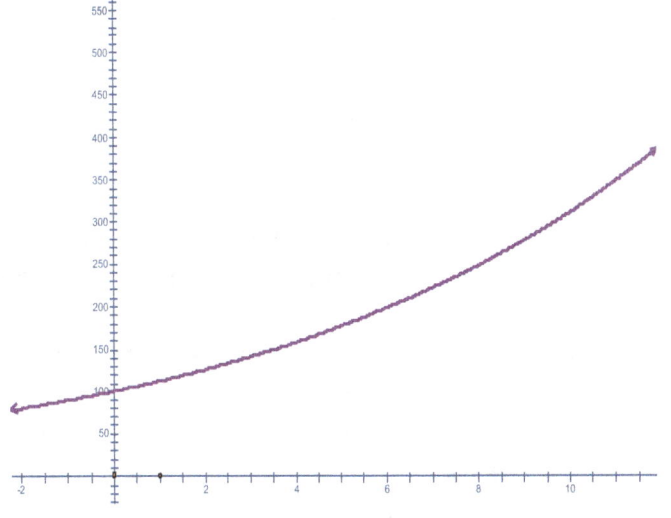

We note that the domain of the algebraic model is bigger than that of the real-life situation. The function $y = 100 \cdot (1.12)^x$ has a domain of all real numbers, since we could input any real number x into this function, including negative values. And the range of the function would be all numbers $y > 0$, since we can get out any positive value as an output from this function.

However, given the context of this problem (in terms of investing money), we are really only interested in the piece of the graph where $x \geq 0$, since we can't go backwards in time, so it would make sense to say the domain of this bank account function would be all times $x \geq 0$, and then the range would be all dollar amounts $y \geq 100$.

We could go even further to note that in the real world, we are likely only interested in amounts of money rounded to the nearest penny. The real-life situation has discrete outputs (money to the nearest cent) whereas the theoretical model has continuous outputs (all positive real numbers form a continuum).

In all of our discussion about this bank account example, we have been talking about the function between the amount of money in the account and the time. But does this really fit the definition of a function? *Why is this a function?* The reason is not that we can write a formula. The reason is not because we can make a graph. None of these are part of the definition of a function. *Go back and read the definition again.* The reason is that for each length of time, there is exactly one amount of money in the account. In order for this relationship to *not* be a function, there'd have to be a time for which there are two different amounts in the account, or a time when there is no amount in the account. (We don't mean 0 dollars, since 0 is an amount, we mean no value at all, not even zero). Since for each time there is one and only one amount of money in the account, we can say the amount of money is a function of time.

So what's the big deal with functions? Why do mathematicians like them so much? Here's the secret: functions are powerful models for many real-life phenomena. So if we can look at something that naturally has a relationship in which there is only one y for each x, then we might be able to capture that functional relationship *algebraically* (by this we mean *with a formula*) and then we can study the phenomenon by studying the algebraic model for the relationship. *Read this paragraph again.* It's a big deal.

Homework Set 18

1) Give some reasons why the relationship between a domain of the set of your classmates and a range of the set of their favorite colors may not be a function.

2) Revisit the equations in the "Get Coordinated" Class Activity in a previous section. For each equation, determine if y is a function of x. Explain why or why not. If y is a function of x, specify all the real numbers x that are in the domain and all the real numbers y that are in the range.
    a) $3x + 4y = 12$
    b) $x^2 + y^3 = 1$

3) Revisit the equations you graphed in the homework in a previous section. For each equation, determine if y is a function of x. Explain why or why not. If y is a function of x, specify all the real numbers x that are in the domain and all the real numbers y that are in the range.
    a) $x^2 + y = {}^-6$
    b) $x^2 + y^2 = 25$
    c) $y^2 = x^3 + 3x^2$
    d) $y = |x|$

4) Sketch the graph of the equation. Use graph paper and show the graph with x and y between $^-10$ and 10. Find and label enough points on the graph to reveal the overall shape of the graph. If the equation describes y as a function of x, specify which real numbers are its domain and which real numbers are in its range. If the equation does not describe y as a function of x, prove that it is not a function.

   a) $\frac{1}{2}x = 3y - 5$

   b) $x + y^2 = 9$

   c) $2x^2 + y^3 = 8$

   d) $x^4 = x^2 + y^2$

5) Consider the Handshake Problem again. In the Read and Study, we claimed there is a functional relationship between the number of members of the club and the number of handshakes that take place. Specify exactly which real numbers are in the domain of this function, and which real numbers are in its range.

6) For each function described below, do the following:
   - make an accurate sketch of its graph on grid paper.
     Find and label enough points to make clear the overall shape of the graph
   - Write a formula for the function using function notation
   - Specify the domain and range of the function

   a) Let $g(x)$ be the function that: squares, add 4, takes the square root.

   b) Let $h(x)$ be the function that: takes the square root, adds 4, squares.

   c) Let $k(x)$ be the function that: adds 4, takes the square root, squares.

   d) Let $m(x)$ be the function that: adds 4, squares, takes the square root.

7) For each function formula, make a machine diagram showing the sequence of operations done to x. Then determine the domain and range for the function.

   a) $f(x) = \sqrt{\frac{1}{x} + 3}$

   b) $f(x) = \frac{1}{\sqrt{x+3}}$

   c) $f(x) = 2x^3 - 5$

   d) $f(x) = (2x - 5)^3$

   e) $f(x) = |x - 3| + 3$

   f) $f(x) = |x + 3| - 3$

   g) $f(x) = {}^-4 \cdot \left(\frac{1}{3x} + 2\right)$.

Class Activity 19a: Bridge Trusses

A common way of constructing bridge trusses is shown below. For example, the truss that is 3 units long is constructed from 11 unit-long beams. Your job is to determine the number of beams required to make a truss that is n units long.

1 unit long   2 units long   3 units long

a) How many beams are required to make a truss that is that is 50 units long?

b) Explain why the number of beams that are required to make a truss that is n units long is a function of n. Specify the domain and range of this function.

c) Write a function formula for the number of beams that are required to make a truss that is n units long. Make an argument based on the structure of the trusses to convince me that your formula is correct and will work for every value of n.

d) Write another function formula for the number of beams that are required to make a truss that is n units long, that counts the beams in a different way than you did in part a, and make an argument for why this other formula makes sense based on the structure of the trusses.

Class Activity 19b: Graphing Relationships

Sketch a graph that could represent the given situation. Make sure that you label and provide a consistent scale on each axis. Be prepared to defend why you think your graph is reasonable.

Then decide whether each scenario behaves like a function. Use the definition of a function to justify your answer. If the scenario behaves like a function, which set is the domain and which set is the range? Would the scenario still behave like a function if you switched which sets you considered the domain and range? Explain.

a) A poorly paced distance runner running a 10 km race starts off quickly, tires and slows to a walk, then sprints to the finish. Sketch a graph showing the runner's total distance traveled over time.

b) A swimmer swims a 200-meter race in a 50-meter pool. She dives from the starting block, and then swims laps in her lane. She swims at a constant rate, except that she has to rest for a few seconds at 150 meters. Sketch a graph of the distance from the starting block over time.

c) Sketch a graph showing the relationship between the **horizontal** distance that you think a cannonball would travel based on the angle of elevation at which it is launched. Show angles from zero degrees (horizontal to the ground) to 90 degrees (perpendicular to the ground).

d) You plan to have fresh fruit pies at your graduation party. Sketch a graph showing the relationship between the number of people you invite to your graduation party and how many pies you think you should bake.

Read and Study 19: Functional Thinking

An idea is always a generalization, and generalization is a property of thinking. To generalize means to think.

George Hegel

In this section, we further explore the idea that algebra is generalized arithmetic that we discussed in the first section of this text. If you truly solved the Handshakes problem in that first section, you didn't just find the number of handshakes problem for one specific number of people, you were able to solve the problem for **any** number of people. This is really amazing if you think about it: there are an infinite number of possible sizes for the club, so there are an infinite number of handshake problems. But the power of mathematical thinking is that we treat all of these possibilities as a single problem, and in that way we can solve an infinite number of problems all at once.

Solving the Handshake problem is really the same thing as finding a general formula for the number of handshakes as a function of the number of club members. The key to doing this is paying attention not to the specific numbers, but to the processes involved and the inherent structure of the problem, and by writing down how the number of trips can be computed in the general case.

Mark Driscoll, in his book *Fostering Algebraic Thinking,* discusses several "algebraic habits of mind". One is to focus on asking **"What changes? What stays the same?"**. These questions are key to identifying a functional relationship between variables in pattern generalization problems. When finding an algebraic formula that fits a numerical pattern, are two important ways of thinking that we'd like highlight about how things are changing and staying the same: **Recursive Thinking**, and **Functional Thinking**. Let's use the Handshakes problem as our example to show these two ways of thinking in action. Here is a table showing our preliminary data.

Number of people	Number of Handshakes
2	1
3	3
4	6
5	10

Recall that we came up with two different ways of solving this problem. One way was to focus on how the number of handshakes for n people depends on the number of handshakes for the previous number of people. For example, there 10 handshakes for 5 people. If we think about how the number of handshakes would change if we added a 6th person, we can realize that there should be 5 more handshakes, since that 6th person would have to shake hands with the 5 previous people. This is an example of recursive thinking. **Recursive thinking** is relating a value (such as the number of handshakes) to its previous value. In the table on the left, we write how number of handshakes come from the previous number of handshakes.

Number of people	Number of Handshakes
2	1
3	$3 = 1 + 2$
4	$6 = 1 + 2 + 3$
5	$10 = 1 + 2 + 3 + 4$
6	$15 = 1 + 2 + 3 + 4 + 5$
n	$1 + 2 + 3 + \cdots + (n-1)$.

Number of people	Number of Handshakes
2	$1 = \frac{2 \cdot 1}{2}$
3	$3 = \frac{3 \cdot 2}{2}$
4	$6 = \frac{4 \cdot 3}{2}$
5	$10 = \frac{5 \cdot 4}{2}$
6	$15 = \frac{6 \cdot 5}{2}$
n	$\frac{n \cdot (n-1)}{2}$

Our second way of solving the problem is shown in the table on the right. This relates the number of handshakes to the number of people n, rather than the previous number of handshakes. **Functional thinking** is relating a value (such as the number of handshakes) to the value of another variable (such as the number of people). For example, we could argue that if there were 7 people, and we asked each of these 7 people how many hands they shook, they would each say 6 hands. So a total of $7 \cdot 6$ handshakes would be reported. However, since each handshake was reported by two people, the total number of handshakes is $\frac{7 \cdot 6}{2}$. In general, we could write that for n people, the number of handshakes is $\frac{n \cdot (n-1)}{2}$.

In our experience, students tend see recursive relationships easier, and struggle in finding functional relationships. However, functional formulas are often much more powerful than recursive ones. For example, the recursive solution we found was that for n people there will be $1 + 2 + 3 + 4 + \cdots + (n-1)$ handshakes. While this solution is correct and makes sense, it does have a drawback in that it is not very practical. In order to find the number of handshakes for, say, 100 people, we'd have to add up the numbers $1 + 2 + 3 + 4 + \cdots + 99$. (So really, to solve the problem with 100 people, we first need to solve the problem for 99 people, which means we first need to solve the problem for 98 people, and so on.) Whereas, with the functional formula, to find the number of handshakes for 100 people, we can just substitute the number 100 and calculate $\frac{100 \cdot (99)}{2}$.

However, recursive thinking can be very powerful, and in fact can be used as a starting point to finding functional formulas as well. We will illustrate this with the Bridge Truss problem. Hopefully in class you and your classmates found several ways of thinking about how to count the number of beams as the length of the bridge grows. Likely, one of those ways was **recursive**, recognizing that to make a truss that is one unit longer, you need to add four more beams to the previous truss. A length one Truss uses 3 beams, then each longer truss uses 4 more beams. We will show this recursive relationship in a table, then discuss how to turn that recursive relationship into a functional one.

Length of Truss	Number of Beams
1	3
2	$7 = 3 + 4$
3	$11 = 7 + 4$
4	$15 = 11 + 4$
5	$19 = 15 + 4$

Length of Truss	Number of Beams
1	3
2	$3 + 4$
3	$3 + 4 + 4$
4	$3 + 4 + 4 + 4$
5	$3 + 4 + 4 + 4 + 4$

Above are two tables showing this recursive relationship in two different ways. Let's think about what the number of beams would be for a truss of length 6 using the **"What Changes? What Stays the Same?"** habit of mind. In the table on the left, what *stays the same* is that we always add 4 more beams. What *changes* is the previous number of beams.

But to write a functional formula, we want to relate the number of beams not to the previous number of beams (that'd be recursive) but to the length of the truss. Relating the number of beams to the length of the truss is functional thinking. If we try to do this in the table on the left, nothing in the number of beams column seems to be related to those numbers 1, 2, 3, 4, 5 for the lengths of the trusses. The fact that the "previous number of beams" is always changing to different numbers is annoying when trying to write a formula. So instead, let's go back to how these numbers are generated, and **focus more on what is staying the same.** For example, instead of saying that truss of length 3 has 7 + 4 beams, let's use the fact that that 7 came from the 3 + 4 beams for the previous length truss.

The result is that in the table on the right we can see that what *stays the same* is the starting value of 3 beams and that we are always adding 4 more beams. What *changes* is the number of times we add the 4. But now we are in position to start thinking functionally. All that is changing is the number of times we add 4. How does the number of times we add 4 depend on the length of the truss? Counting the number of times we add 4, and writing this repeated addition as multiplication, we get

Length of Truss	Number of Beams
1	$3 + 4 \cdot 0$
2	$3 + 4 \cdot 1$
3	$3 + 4 \cdot 2$
4	$3 + 4 \cdot 3$
5	$3 + 4 \cdot 4$

Here we can see that all that is changing is the number we are multiplying 4 by, and moreover, it is easy to see that this number is one less than the length of the truss. (Why does that make sense?) So a bridge with length 10 will need $3 + 4 \cdot 9$ beams, and in general, a bridge with length n will need $3 + 4 \cdot (n - 1)$ beams.

So we have taken a recursive pattern (adding 4 more beams to make the next truss) and turned it into a functional formula (starting with 3 beams and adding 4 a total of $n - 1$ times). Hopefully in class, you or your classmates found other ways of solving this problem, leading to different formulas. We will ask you about some other possibilities in the homework.

Homework Set 19

1) Shown below are several formulas representing a solution to the Bridge Truss problem. Justify each formula by explaining how it counts the number of beams, based on the structure of the truss.

   a) $f(n) = n + (n - 1) + n + n$

   b) $f(n) = 3 \cdot n + (n - 1)$

   c) $f(n) = 4 \cdot n - 1$

2) Consider the pattern shown below. The square grid that is 3 units long is made from 24 unit-long segments.

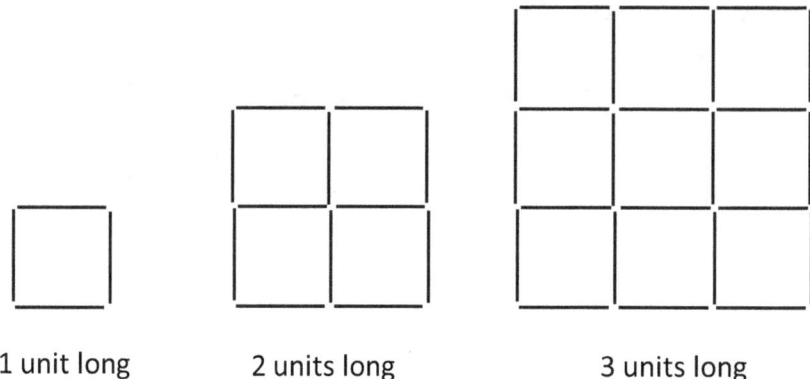

1 unit long   2 units long   3 units long

   a) How many unit long segments are in the figure that is 4 units long? How many are in the figure that is 20 units long?

   b) Write a function formula for the number unit long segments are in the figure that is n units long. Make an argument based on the structure of pattern to convince me that your formula is correct and will work for every value of n.

3) Bob's Discount offered me the following deal: I pay $25 for a membership card, and then I get 10% off of every purchase I make during the next year.
   a) How much in total do I need to purchase at Bob's to break even on this deal?
   b) If I purchase the amount in part a) during the next year, consider my "net savings" to be $0. Make a table showing my net savings with the card as a function of total amount purchased during the next year.
   c) Make a graph of this function.
   d) Write an algebraic formula for this function.
   e) What is the domain of this function? What is the range?

4) A senator has proposed the following income tax plan: an individual would pay no taxes on their first $10,000 income, 5% of their next $10,000, 10% of their next $10,000, 15% of their next $10,000, 20% of their next $10,000, and so on, until $100,000, after which all income is taxed at 50%. For example, an individual that makes $50,000 per year would be taxed 0 + 500 + 1000 + 1500 + 2000 for a total of $5000.
   a) Make a table showing the amount of taxes paid as a function of income. Show incomes up to $200,000.
   b) Sketch a graph of the amount of taxes paid as a function of income.
   c) Use the definition of a function to explain why the total amount of taxes paid is a function of income.
   d) Give a clear description of the domain and the range of this function, and state precisely which real numbers are possible in the domain and which real numbers are possible in the range.

Class Activity 20: Colony Data

In the year 2000, ($t = 0$), three spaceships of scientists were sent to explore and colonize three different planets in the Gamma Quadrant. Here are data for each colony in the first decades:

Colony I		Colony II		Colony III	
t (decades)	Population	t (decades)	Population	t (decades)	Population
1	56	1	279	1	58
2	70	2	543	2	90
3	88	3	807	3	146
4	110	4	1071	4	226
5	137	5	1335	5	330
6	172	6	1599	6	458

1) What was the initial population (at time $t = 0$) of each colony? Predict the population at time $t = 7$ for each colony.

2) Determine which colony's population data is best modeled by a **linear function** of the form $P(t) = at + b$.
    - How can you recognize linear growth *from the table*?
    - Figure out values for the parameterss a and b so that the function $P(t) = at + b$ fits the data for that colony.
    - Try to interpret the meaning of each constant in your model.

3) Determine which colony's population data is best modeled by an **exponential function** of the form $P(t) = a \cdot b^t$.
    - How can you recognize exponential growth *from the table*?
    - Figure out values for the parameters a and b so that the function $P(t) = a \cdot b^t$ fits the data for that colony.
    - Try to interpret the meaning of each constant in your model.

4) Determine which colony's population data is best modeled by a **quadratic function** of the form $P(t) = at^2 + bt + c$

    - How can you recognize quadratic growth *from the table*?
    - Figure out values for the parameters a, b and c so that the function $P(t) = at^2 + bt + c$ fits the data for that colony.
    - Try to interpret the meaning of each constant in your model.

5) Order the Colonies from largest to smallest based on what their populations will be in the year 2200 ($t = 20$) assuming that growth continues in this way.

Also in the year 2000, (t = 0), three other spaceships of scientists were sent to explore and colonize three different planets in the Delta Quadrant. Unfortunately, these colonies were not successful. Here are population data for each of these colonies in the first decades:

Colony IV			Colony V			Colony VI	
t (decades)	Population		t (decades)	Population		t (decades)	Population
1	520		1	520		1	520
2	420		2	338		2	440
3	330		3	220		3	360
4	250		4	143		4	280
5	180		5	93		5	200
6	120		6	60		6	120

1) What was the initial population (at time $t = 0$) of each colony? Predict the population at time $t = 7$ for each colony.

2) Determine which colony's population data is best modeled by a **linear function** of the form $P(t) = at + b$.
   - How can you recognize linear growth *from the table*?
   - Figure out values for the parameters a and b so that the function $P(t) = at + b$ fits the data for that colony.
   - Try to interpret the meaning of each constant in your model.

3) Determine which colony's population data is best modeled by an **exponential function** of the form $P(t) = a \cdot b^t$.
   - How can you recognize exponential growth *from the table*?
   - Figure out values for the parameters a and b so that the function $P(t) = a \cdot b^t$ fits the data for that colony.
   - Try to interpret the meaning of each constant in your model.

4) Determine which colony's population data is best modeled by a **quadratic function** of the form $P(t) = at^2 + bt + c$

   - How can you recognize quadratic growth *from the table*?
   - Figure out values for the parameters a, b and c so that the function $P(t) = at^2 + bt + c$ fits the data for that colony.
   - Try to interpret the meaning of each constant in your model.

5) Which Colony will die out first?

Read and Study 20: Linear, Exponential and Quadratic Growth

Your pain is the breaking of the shell that encloses your understanding.
Kahil Gibran

Functions are all about change. If we change the input, we change the output. In Table I from the Class Activity, we can see a pattern in how *y* changes with respect to *t*. Namely, as *t* increases by one, the population increases by *multiplying* by 1.25 (or we could say *P* increases by a *factor* of 1.25). This pattern holds throughout the entire table.

This kind of change, where for a fixed increase in the independent variable results in *multiplying* the dependent variable by constant factor, is called **exponential change**. Here there are constant **ratios** between successive *y* values for constant changes in *x*. Functions that have this pattern can be modeled using an **exponential form**. For example, $f(x) = a \cdot b^x$ is a standard exponential form. Here *b* is called the base of the exponential function. To model a function that changes exponentially, we can figure out the appropriate *a* and *b* and then use the model $f(x) = a \cdot b^x$ to answer questions about the specific situation.

In contrast, with **linear change**, a fixed increase in the independent variable results in *adding* a constant amount to the dependent variable. Linear change is characterized by constant **differences** in population for constant changes in time. The appropriate model is a **linear form**, such as $f(x) = ax + b$. For example, in the Bridge Truss problem, the number of beams was a linear function of the length of the truss, since increasing the length of the truss by one always resulted in adding 4 beams.

It is very informative to look at how similar exponential and linear functions are. In both cases, the function can be thought of as changing at a constant rate. The key is that in exponential functions this is a constant *multiplicative* rate, while for linear functions this is a constant *additive* rate. This similarity shows up in the standard function forms as well. In the standard linear form $f(x) = ax + b$, the slope *a* represents the constant additive rate of change and the *y*-intercept *b* is the "initial" value of the function when *x* is zero. In other words, this function starts out with a value of *b*, then *adds* on *a* a total of *x* times. In the standard exponential form $f(x) = a \cdot b^x$, the base *b* represents the constant multiplicative rate of change, and the *y* intercept *a* is the "initial" value of the function when *x* is zero. In other words, this function starts out with a value of *a*, then *multiplies* by *b* a total of *x* times.

Sometimes you can see a pattern where the dependent variable does not change linearly (by adding a constant amount), but by adding an amount that itself grows linearly. This is the pattern that characterizes **quadratic growth**. Here, there are constant **second differences** in the dependent variable. Make sure to talk about this in class. Quadratic growth can be modeled by a function in **quadratic form**, such as $f(x) = ax^2 + bx + c$ or $f(x) = a(x - h)^2 - k$. In later sections we will explore how changes in the parameters *a*, *k*, and *h* change the graph of a quadratic function.

Hopefully in the Class Activity, you figured out the population for Colony III is best modeled by a quadratic function. Let's discuss how one way that we can figure out what values for the parameters *a*, *b* and *c* so that the function $P(t) = at^2 + bt + c$ fits the data for that colony. And hopefully you also figured out that the initial population for this colony was 50 people. In other words $P(0) = 50$. If we substitute $t = 0$ into our function form, we'd get

$$P(0) = a \cdot (0)^2 + b \cdot (0) + c \qquad \text{Substitute } t = 0$$

$$P(0) = c \qquad \text{Simplify right hand side}$$

$$50 = c \qquad \text{Since } P(0) = 50$$

So the parameter *c* must be the initial population, and to fit our data, we must have $c = 50$. So now we have our function looking like this:

$$P(t) = at^2 + bt + 50$$

To figure out what *a* and *b* need to be, we can continue to substitute in more data points. According to the table, we have $P(1) = 58$. Substituting into our function form we have

$$P(1) = a \cdot (1)^2 + b \cdot (1) + 50 \qquad \text{Substitute } t = 1$$

$$P(1) = a + b + 50 \qquad \text{Simplify right hand side}$$

$$58 = a + b + 50 \qquad \text{Since } P(1) = 58$$

$$8 = a + b \qquad \text{Subtract 50 from both sides}$$

Unfortunately, this didn't tell us what either *a* or *b* has to be, but at least we know they must add to 8. Let's substitute in another data point from the table, say $P(2) = 90$.

$$P(2) = a \cdot (2)^2 + b \cdot (2) + 50 \qquad \text{Substitute } t = 2$$

$$P(2) = 4a + 2b + 50 \qquad \text{Simplify right hand side}$$

$$90 = 4a + 2b + 50 \qquad \text{Since } P(2) = 90$$

$$40 = 4a + 2b \qquad \text{Subtract 50 from both sides}$$

So now we another linear equation for a and b. So we can solve this system of two linear equations:

$$8 = a + b$$
$$40 = 4a + 2b$$

to find out what a and b must be. We'll let you use your favorite method to do this (substitution, elimination, using algebra tiles on balance scales...). We found that $a = 12$ and $b = {}^-4$. So using these values in our function form, we get

$$P(t) = 12t^2 - 4t + 50$$

as the quadratic function that models the population in Colony III. We know for sure that this fits the populations at times 0, 1, and 2, since we used those populations to find the values for the parameters, but we encourage you to check to see that this function formula also generates the other populations at times 3, 4, 5 and 6.

Of course not all data in the world fits one of these three function forms we've described in this section. We will also explore some higher order polynomial functions, rational functions and logarithmic functions in this text. Trigonometric functions, (such as sine, cosine and tangent) are great for modeling data that oscillates, but they are the subject of another course.

Homework Set 20

1) The table below shows the number of rabbits living in a field over the course of several months. The number of rabbits listed is what was observed on the first day of each month.

Month	Number of Rabbits
April	230
May	278
June	337
July	407
August	500

   a) At what *rate* is the population of rabbits growing? Explain.
   b) Find an algebraic model that fits this data.
   c) If next year the same field starts with only 40 rabbits on April 1, how many would you expect there to be by August 1?

2) A child launches a "stomp rocket" from the ground up into the air; it reaches a maximum height of 64 feet after 2 seconds before it comes back down to the ground (after 2 more

seconds). Model the position of the ball above the ground as a function of time using a quadratic form. Carefully explain the meaning of each parameter in your function.

3) Here are the data for growth in three rabbit colonies given in years. In each case...
   a) Sketch a graph of the population as a function of time;
   b) Fit either a linear, quadratic, or exponential model to each data set, and write a formula for the population as function of time in years.
   c) Use your model to find the initial population (Year 0) and the population in year 50.

Colony A		Colony B		Colony C	
Year	Pop.	Year	Pop.	Year	Pop.
5	1256	5	560	5	800
10	1287	10	542	10	900
15	1319	15	524	15	900
20	1353	20	506	20	800
25	1387	25	488	25	600
30	1422	30	470	30	300

4) Recall the square grid problem from the previous section.

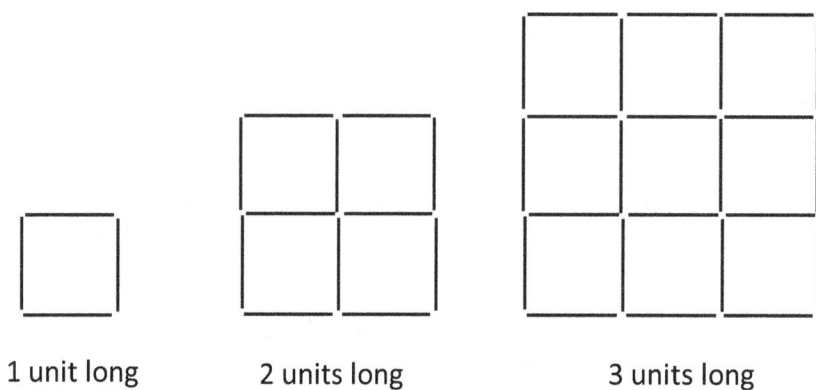

1 unit long   2 units long   3 units long

a) Make a table that shows the number of segments as a function of the length, and compute the second differences to show that they are constant.
b) Write the function formula for the number of segments in the standard quadratic form $f(x) = ax^2 + bx + c$.

5) Recall the handshake problem discussed again in the previous section.
a) Make a table that shows the number of handshakes as function of the number of people, and compute the second differences to show that they are constant.
b) Write the function formula for the number of handshakes in the standard quadratic form $f(x) = ax^2 + bx + c$.

Class Activity 21: Crossing the River

One day two children and some adults went hiking through the woods. While hiking, they came upon a large river (unsuitable for swimming) that they all wanted to cross. There was no bridge, but they happened to find a small abandoned boat and oars. They decided that the canoe could safely transport both children, but only one adult at a time (with no children). What is the least number of times the canoe must cross the river in order to get all of them safely across? (Yes, a child is able to row the boat alone. No, the canoe cannot be sent back with no one in it.)

a) Suppose 25 adults are in the group. What is the minimum number of trips required?
b) Make a graph showing the relationship between the minimum number of trips required and the number of adults.
c) Is it appropriate to connect the dots on your graph? Why or why not?
d) Is the minimum number of trips required a function of the number of adults in the group? Explain.
e) Write a formula for the number of trips required to get a adults across the river. Interpret the meaning for each number in your formula.

Read and Study 21: Linear Function Forms

Your pain is the breaking of the shell that encloses your understanding.
                                                                                                  Kahil Gibran

As we discussed in the first chapter, **modeling a problem algebraically** means finding an algebraic (or sometimes a numeric or graphic) representation of a function that captures the essential elements of the situation. Mathematicians have made up names and notations to represent many types of functional relationships; we'll call these **standard function forms**. If a problem can be modeled by a standard type, we can go ahead and use the appropriate form and fill in the constants to fit that exact situation.

Functions are all about change. If we change the input, we change the output. (Except for constant functions, in which the output is constant no matter what the input is. But this is kind of a boring type of function. So let's just say that all *interesting* functions are all about change.) Often, if we are lucky, we can see a pattern in the way that a function is changing. In the Crossing the River Problem, you found that as the number of adults in the group increases, so does the number of trips required. More precisely, we found that increasing the number of adults by 1, results in an increase of 4 more required trips.

This type of relationship is called a linear function, with a constant rate of change of 4 required trips per 1 adult. The function formula you found was likely $f(x) = 4x + 1$, which has a graph that is a straight line with slope 4 and y intercept at (0,1). You also should have found that the inverse function was also a linear one, though naturally it had different values for the slope and y-intercept. In general linear function can be written in the form $f(x) = mx + b$. This form that shows that its graph will be a line with slope m and y intercept $(0, b)$.

Equivalent Algebraic Forms for Functions. Typically, there are many different equivalent forms for a given function, and each form can make different features of the function or its graph more apparent. As an illustration of this, we will review the different forms that a linear function might take. A linear function has the simplest of graphs: the straight line. There are many different forms in which you can write the equation for a line, depending on what features you want to highlight, or depending on which information about the line you have handy to plug in.

All these forms may look different in the table that follows may look different, but have in common is that there is a term that involves a constant times x, a term that involves a constant times y, and a constant term. In particular, there are never any terms with x times y, or x or y to any power other than 1. (Equations of this type are called linear, since their graphs are straight lines.)

Common Forms of Equations for a line with variables x and y.

Form Name	Equation	Graphical Interpretation of Parameters
Standard	$ax + by = c$	None, individually. See Problem Set
Slope-Intercept	$y = mx + b$	m is the slope, b is the $y-$intercept
Point-Slope	$y = m(x - h) + k$	m is the slope, (h, k) is a point on the line
Two-Point	$y = \left(\dfrac{k_2 - k_1}{h_2 - h_1}\right)(x - h_1) + k_1$	(h_1, k_1) and (h_2, k_2) are points on the line
Two Equidistant Points	See Problem Set	(a_1, b_1) and (a_2, b_2) are two points equidistant to the line.

Moreover, for any particular line, all of these different forms would be equivalent equations. Sometimes, you may have an equation in one form, and want to rewrite it in a different form, depending on which information about the line you want to make clear. For example, suppose you have the equation for a line written in "standard form", such as $2x + 5y = 30$, and wanted to write it in "point-slope" form. One way would be to find two points on the line, used those points to find its slope, and then substitute this information into the "point-slope" form of a line. Let's do that now. Here's a table of points we generated from the equation, and then a graph we made using those points.

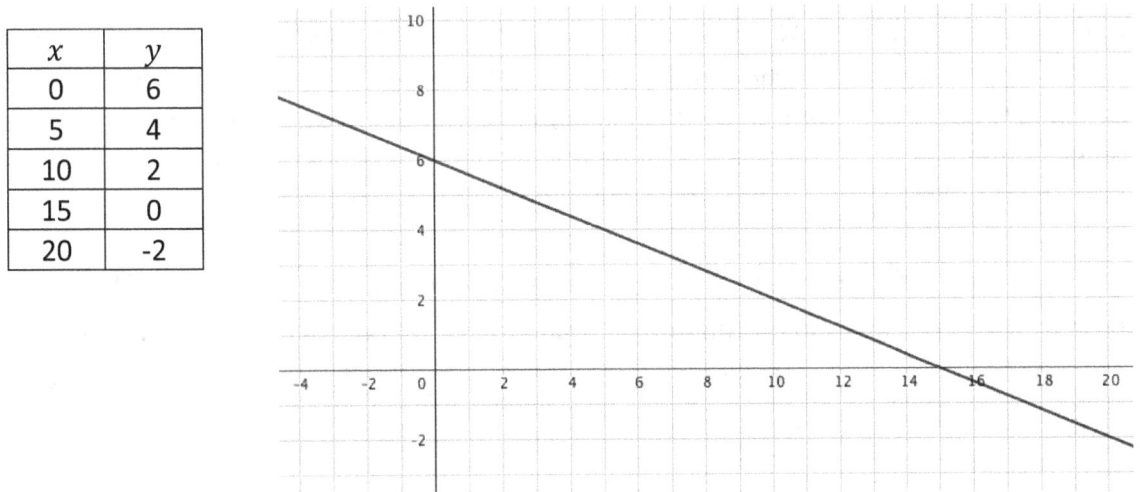

x	y
0	6
5	4
10	2
15	0
20	-2

Using any two points, or by reasoning from the graph, we see that the slope of this line is $-\dfrac{2}{5}$, meaning the y value decreases by 2 whenever the x increase by 5. For fun, let's write the equation for this line in "point-slope" form, using the point $(5,4)$. Substituting the information

157

we have into the "point-slope" form, we'd get $y - 4 = -\frac{2}{5}(x - 5)$. (Stop and make sure you confirm this by doing it yourself.)

Now here's our main point: the equations $2x + 5y = 30$ and $y - 4 = -\frac{2}{5}(x - 5)$ are equivalent. They have the same set of solutions. We trust that they are equivalent, since we know they both have the same graph (which is the set of solutions!). But we can also show they are equivalent by using the definitions and laws of algebra. We will start with the first equation, then write a sequence of equations, each of which we can justify as being equivalent to the previous.

$$2x + 5y = 30$$

$$5y = {}^-2x + 30 \qquad \text{Property of Equality: Add } {}^-2x \text{ to each side}$$

$$y = \frac{{}^-2x + 30}{5} \qquad \text{Property of Equality: Multiply } \frac{1}{5} \text{ to each side}$$

$$y = -\frac{2}{5}x + 6 \qquad \text{Distributive Law: } \frac{1}{5} \cdot ({}^-2x + 30) = \frac{1}{5} \cdot ({}^-2x) + \frac{1}{5} \cdot 30$$

$$y = \left(-\frac{2}{5}x + 2\right) + 4 \qquad \text{Since } 6 = 2 + 4$$

$$y = -\frac{2}{5}(x - 5) + 4 \qquad \text{Distributive Law: factor out } -\frac{2}{5} \text{ from } -\frac{2}{5}x + 2$$

If we leave it in the form of the fourth equation, $y = -\frac{2}{5}x + 6$, that's the "Slope-Intercept" form. The reason we chose in the fifth equation to split up the 6 into $2 + 4$ was to write the equation in "Point-Slope" form using the point where the y value was 4. If we wanted to write it using another point, we could have split up the 6 differently. Note that the "Slope-Intercept" form is just the "Point-Slope" form where the point used is the y-intercept.

Now suppose instead at the beginning we were given the equation $y = -\frac{2}{5}(x - 5) + 4$, and wanted to write the equation for this line in "standard form", we could do the same sequence of equations as above, but in reverse, and get that $y = -\frac{2}{5}(x - 5) + 4$ is equivalent to $2x + 5y = 30$.

Lastly, we said that in standard form the parameters a, b, and c don't individually have a graphical interpretation. But notice that in our example, a is 2 and b is 5, and the slope was $-\frac{2}{5}$. That's not a coincidence, as we will ask you to show in the homework.

In a previous class activity, we determined that the set of points that is equidistant between two given points is a straight line, so we included in our table a spot for a "Two Equidistant Points" form of a line. We'll admit that giving two points that are equidistant to the line is not a common way to define a line. But it works. And who knows, if someday you find that have an application where you want to find the equation for the line given two equidistant points it passes between (see the problem set for some) then you might find it handy to write the equation in that form instead of a slope intercept form, since the slope of the line might not reveal anything of use to you.

The big idea is that there are different algebraic forms for the same function, and different algebraic forms an equation of the same graph, and the most useful one for your situation depends on the context. Being proficient in algebra means being flexible and being able to take an equation in one form and using the definitions and laws of algebra to rewrite it into an equivalent equation in a different form.

Homework Set 21

1) In the read and study we investigated the line defined by the equation $2x + 5y = 30$. Write a story problem that would result in this equation, that gives meaning the numbers 2, 5 and 30, and to the variables x and y.

2) Consider the line with equation $y = 3x - 5$.
   a) Sketch a graph of this line.
   b) Write the equation for this line in "standard form".
   c) Write the equation for this line in "point-slope" form.

3) Consider the line with equation $4x - 2y = 7$
   a) Sketch a graph of this line.
   b) Write the equation for this line in "slope-intercept" form.
   c) Write the equation for this line in "point-slope" form.

4) Suppose postage for package is $5 for the first pound and $2 for each additional pound. Model the postage price as a function of weight using a linear form. Carefully explain the meaning of each variable in your function.

5) In the "standard" form of a line $ax + by = c$, we said the parameters a, b, and c don't have ready graphical interpretations. That means that, for example, the a by itself doesn't tell you anything about the graph. But in combinations, they can.
   a) Find an expression for the slope of the line given by $ax + by = c$.
   b) Find an expression for the x-intercept of the line $ax + by = c$.
   c) Find an expression for the y-intercept of the line $ax + by = c$.

6) Consider the following scenario: A cell phone company offers two monthly data roaming plans. Plan A (Pay As You Go) costs $0.10 per MB. Plan B (Data Bundle) costs $20 for 1 Gig of data valid for that month. If you use more than 1 Gig that month, any overages will be charged at the Pay as You Go rate of $0.10 per MB. (Note: 1 Gig is 1000 MB).
   a) Make a table showing the relationship between the amount of data used per month and the cost of plan A. Show the cost for every 200 MB of data, up to 2 Gigs.
   b) Make a table showing the relationship between the amount of data used per month and the cost of plan B. Show the cost for every 200 MB of data, up to 2 Gigs.
   c) Using grid paper, make a graph of the cost for Plan A and a graph for the cost of Plan B on the same coordinate axes. The costs should be on the vertical axis and the data usage on the horizontal axis.
   d) Is it appropriate to connect the dots in the graphs for each plan? Why?
   e) Write a formula for the cost of plan A depending on the number of MB you use. Now Write a formula for the cost of plan B depending on the number of MB you use. Interpret the meaning of each number in your formulas.
   f) Suppose you work for this cell phone company and a customer wants to know which plan to get. What is your advice?

7) In a previous homework problem, you wrote an equation for the line that is equidistant between the points $(-6, 2)$ and $(4, 8)$. Show how you can re-write that equation into standard form and into slope-intercept form.

8) Challenge: Find the equation of the line that passes between (a_1, b_1) and (a_2, b_2) equidistantly. Simplify the equation into standard form. Now use your equation to solve the following problems:
   a) Two competing oil companies each have an oil well located in an unclaimed desert, at coordinates $(3, 5)$ and $(8, 10)$. Your job is to determine a straight boundary line that separates the two oil fields, so that nowhere along that boundary can either company claim that the other company's oil well is closer to the boundary than theirs. Find an equation for your boundary line.
   b) Two lamps, each putting out the same amount of light, are placed at the coordinates $(-2, 5)$ and $(3, -6)$. Write an equation for all the points where the intensity of the light coming from each lamp will be equal.

Class Activity 22: Square Dance

1. The product of two numbers can be interpreted as the area of a rectangle with side lengths given by the two factors. Use Algebra Tiles to make a rectangle with area given by the following expressions. Record a sketch of your rectangle, and write a simplified expression for the total area represented.

    a. $(x + 4)(3x + 2)$

    b. $(x + 4)(3x - 2)$

2. Figure out how to use Algebra Tiles to represent the following expression as the area of a rectangle. Record a sketch of your model.

    $2x^2 + 13x + 6$

3. Discuss whether and how you could use the factorization you found in #2 to help solve the following equations:
    a. $2x^2 + 13x + 6 = 0$

    b. $2x^2 + 13x + 6 = 21$

    c. $2x^2 + 13x + 6 = 12$

4. Use Algebra Tiles to represent the following expression as the area of a square. Record a sketch of your model.
$$x^2 + 10x + 25$$

5. Discuss whether and how you can use the factorization you found in #4 to help solve the following equations:

   a. $x^2 + 10x + 25 = 0$

   b. $x^2 + 10x + 25 = 9$

   c. $x^2 + 10x + 25 = 12$

6. Find solutions to the following equations by first adding or subtracting an amount to both sides so that the left-hand side can be factored as a square. (This method is called "completing the square".)

   a. $x^2 + 10x = 11$

   b. $x^2 + 10x = 5$

   c. $x^2 + 9x + 20 = 0$

   d. $3x^2 + 5x + 2 = 0$

Read and Study 22: Area Models for Quadratic Expressions

Mathematicians do not study objects, but relations among objects; they are indifferent to the replacement of objects by others as long as relations do not change. Matter is not important, only form interests them.

Henri Poincare

The association between arithmetic and geometry is ancient. A number can be thought of as a length, and addition as combining lengths. Areas of rectangles can be found by multiplying the side lengths. Of special interest are squares (rectangles with equal side lengths). The area of a square can be found my multiplying those equal side lengths together. And that is why $x \cdot x$ is commonly referred to as "x squared", originally meaning "take the length of x and make a square with that side length and take its area".

Equations involving only constants, x, and x^2 terms are called **quadratic equations.** The word quadratic refers to the square, ("quad" is Latin for "four", for the four-sided square). Similarly, a **quadratic function** is a function of the form $f(x) = ax^2 + bx + c$, where a, b, and c are real number parameters. (And a isn't zero, or else it would be a linear function instead of quadratic.

There are many commercially available sets of "algebra tiles" for modeling algebra with areas. In this section, we will use red squares representing the unit constant, green pieces representing the variable x and x^2. Negative quantities can be shown by turning any piece "white-side" up. You can print and cut out your own set using the template in the appendix.

The small red squares have side lengths of 1, and so have an area of 1.
The green rectangles have side lengths 1 and x and so have an area of x.
The large green squares have side lengths of x, and so have an area of x^2.

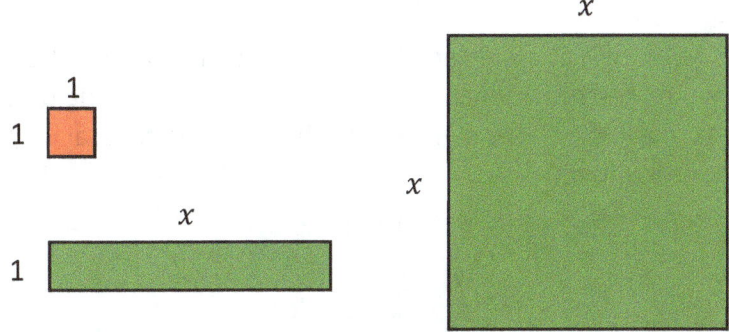

Algebra tiles can be used to model multiplication of algebraic expressions as finding areas of rectangles. For example, the expression $(x + 1) \cdot (2x + 3)$ could be thought of as the area of a rectangle with side lengths $(x + 1)$ and $(2x + 3)$, which could be represented with tiles modeled as shown on the next page.

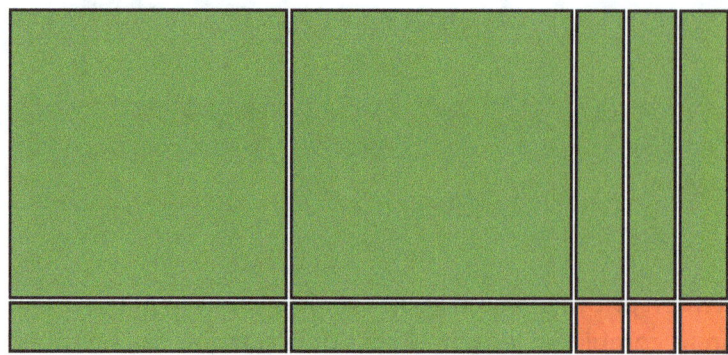

Make sure you can see that the rectangle shown has one side of length $(x + 1)$ and another side of length $(2x + 3)$. The model then can be interpreted to show that $(x + 1) \cdot (2x + 3)$ is equal to $2x^2 + 5x + 3$, as the area of the entire rectangle is composed of two large squares with area of x^2, five long rectangles each with area x, and the three red squares with area 1.

Suppose now you had the following expression: $x^2 + 6x + 8$. This can be represented in tiles as shown:

To factor an expression is to write it as a product. If we could arrange these tiles into a rectangle, then the area would be the product of the two side lengths, giving us a way to factor the expression. This can take a little trial and error, and there can be several ways of doing it, but one possibility would be as shown:

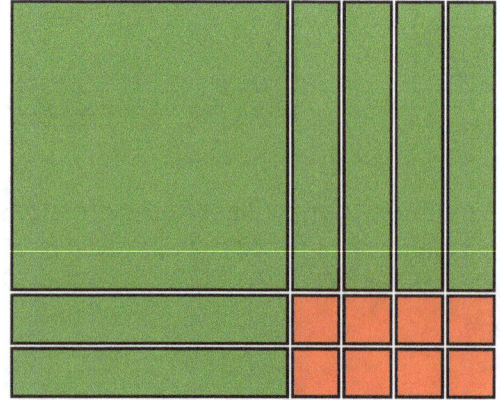

This rectangle has side lengths $(x + 4)$ and $(x + 2)$, and so the area of the rectangle is $(x + 4) \cdot (x + 2)$. So we have two ways of expressing this total area: $x^2 + 6x + 8$ and $(x + 4) \cdot (x + 2)$. So we must have

$$x^2 + 6x + 8 = (x + 4) \cdot (x + 2).$$

The nice thing about being able to factor expressions is that if we can have a factored expression equal to zero, we can use the zero products property to see that one (or both) of the factors must be zero.

For example, suppose we wanted to solve the following equation: $x^2 = -6x - 8$. Then a good approach might be to see if we can write this as a factored expression equaling zero, as follows:

x^2	$=$	$^-6x - 8$	
$x^2 + 6x + 8$	$=$	0	Add $6x + 8$ to both sides
$(x + 4)(x + 2)$	$=$	0	Distributive Law (factor)
$x + 4 = 0$	or	$x + 2 = 0$	Zero Products Property
$x = {}^-4$	or	$x = {}^-2$	Additive Inverses

In the preceding sections, and in the class activity, we also saw how easy it is to solve equations that are of the form

$$(some\ expression)^2 = some\ number,$$

since that equation will be true when the "some expression" is equal to either the square root of the "some number", or negative the square root of that number.

For example, if we want to find solutions to the equation

$$(3x + 4)^2 = 25,$$

we can reason as follows: the expression $3x + 4$ has been squared to get 25. There are only two numbers you can square to get 25, namely 5 and $^-5$. So the expression $3x + 4$ must be one of these. So either

$$3x + 4 = 5, \quad \text{or} \quad 3x + 4 = {}^-5.$$

The first possibility is true if x is $1/3$, the other if x is $^-3$. So those are the two solutions.

Let's use this idea of using squares to find a different way we can use the area model to solve a quadratic equation. Suppose we had the same equation as before:

$$x^2 + 6x + 8 = 0.$$

As we've seen, the expression $x^2 + 6x + 8$ can be represented as this collection of tiles as shown:

Now instead trying to make just any old rectangle, we tried to make that rectangle be a square, we'd find we couldn't do it, we'd be one small red square short.

Since we have six x's, in order for the sides of the square to have the same length, these six x's would have to be split into three and three, so that each side of the square would have to have length $x + 3$. But then
$$(x + 3) \cdot (x + 3) = x^2 + 6x + 9,$$
and we only have $x^2 + 6x + 8$.

But this is not a problem, since we are solving an equation, we can easily add 1 to both sides of the equation. Then one side is the area of a square, and we can solve the equation by thinking about square roots:

$$
\begin{aligned}
x^2 &= {}^-6x - 8 & \\
x^2 + 6x + 8 &= 0 & \text{Add } 6x + 8 \text{ to both sides} \\
x^2 + 6x + 9 &= 1 & \text{Add 1 to both sides (completing the square)} \\
(x + 3)^2 &= 1 & \text{Distributive Law (Factoring)} \\
x + 3 = 1 \quad &\text{or} \quad x + 3 = {}^-1 & \text{+1 and -1 are the numbers you square to get 1} \\
x = {}^-4 \quad &\text{or} \quad x = {}^-2 &
\end{aligned}
$$

Notice how we have now solved the equation $x^2 + 6x + 8 = 0$ in two different ways. In the first example, we expressed $x^2 + 6x + 8$ as the area of a rectangle, and in the second example, we added 1 to both sides to get $x^2 + 6x + 9$ which we could express as the area of a square.

These two standard techniques for solving quadratic equations (usually called "factoring", and "completing the square") each make use of the area model which can be represented with these tiles. Factoring uses rectangles and the Zero Product Property, while completing the square uses squares and the definition of the square root.

Homework Set 22

1) Make an area model (draw algebra tiles) to find the product of $(2x + 3)$ and $(4x + 2)$.

2) Use algebra tiles to factor the following expressions as the area of a rectangle.
   a. $x^2 - 4x - 12$
   b. $2x^2 - 3x + 1$
   c. $6x^2 + 13x + 5$

3) Find solutions to the following equations:
   a. $(2x - 1)^2 = 0$
   b. $(2x - 1)^2 = 16$
   c. $(2x - 1)^2 = 5$

4) Find solutions to the following equations:
   a. $(x + 2)(x + 4) = 0$
   b. $(x + 2)(x + 4) = 16$
   c. $(x + 2)(x + 4) = 5$

5) Show how to use algebra tiles to help you to literally 'complete the square' on the left side of these equation. Then find all solutions to the equation.
   a. $x^2 + 6x = 2$
   b. $x^2 + 8x + 13 = 0$

6) Solve the following equations by the method of completing the square.
   a. $x^2 - 8x = 9$
   b. $5x^2 = 3 + 20x$
   c. $3x^2 + 7x = {}^-2$
   d. (Challenge!) $ax^2 + bx + c = 0$

Class Activity 23a: Form Follows Function

A **quadratic function** is a function of the form $f(x) = ax^2 + bx + c$, where a, b, and c are real number parameters. The form $f(x) = ax^2 + bx + c$ is called the **standard** form for a quadratic function. But as we will now investigate, a rather more convenient form for a quadratic function is to write it in what's called the **vertex** form:

$$f(x) = a(x - h)^2 + k.$$

As an example, consider the quadratic function $f(x) = x^2 - 6x + 4$. Use the method of completing the square to write this function in the vertex form shown above.

We'll get you started. To build on our completing the square skills, let's use the variable y for $f(x)$ to write the quadratic function as an equation, so we can more easily add what we need to both sides to complete the square. Now you figure out how to complete the square. When you are done, solve for y again to write the function using function notation.

$$f(x) = x^2 - 6x + 4$$
$$y = x^2 - 6x + 4 \qquad \text{We let } y \text{ be the dependent (output) variable}$$
$$= \qquad\qquad\qquad\qquad \text{Add \_\_\_\_ to both sides}$$

After you have written the function in the form $f(x) = a(x - h)^2 + k$, do the following:
- Make a table of inputs and outputs and sketch the graph of the function $f(x)$.
- Identify and label the vertex of the graph.
- Find the exact roots of the function.

Which form of the function do you find easiest to use to find these points. Why?

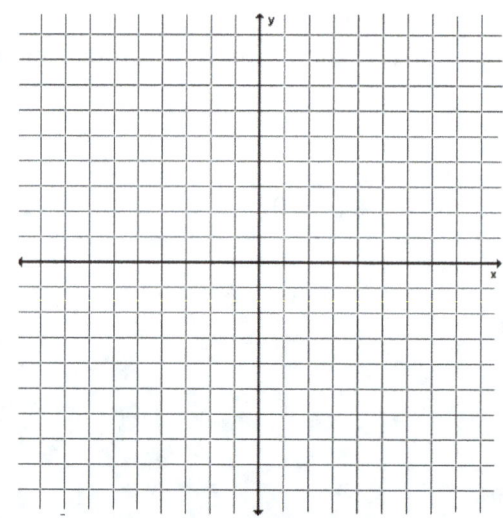

Class Activity 23b: Parabolas and Quadratic Functions

We have already made the following definitions:

- A **parabola** is the set of points equidistant between a point, called the focus, and a line (called the directrix).
- A **quadratic function** is a function of the form $f(x) = ax^2 + bx + c$, where a, b, and c are real number parameters.

Now we will investigate what you've long been told is true: that parabolas are the graphs of quadratic functions.

1. Use the definition of a parabola to write an equation for the parabola with focus $(^-2, 5)$ and directrix $y = 7$, and sketch its graph.

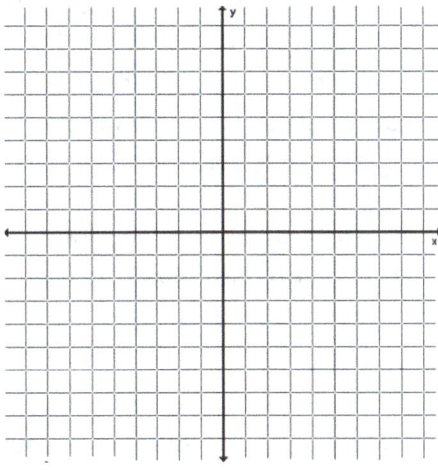

2. Show that this equation can be written in the standard quadratic function form $f(x) = ax^2 + bx + c$.

3. Now show how to rewrite the equation for that parabola in the form
$$f(x) = a(x - h)^2 + k.$$

What is the significance of the h and k in this formula?

Read and Study 23: Quadratic Function Forms

A fact in itself is nothing. It is valuable only for the idea attached to it, or for the proof which it furnishes.

Claude Bernard

A **quadratic function** is a function of the form $f(x) = ax^2 + bx + c$, where a, b, and c are real number parameters. (Here a understood to be non-zero, otherwise it would be a linear function).

You likely already have been told long ago that quadratic functions have graphs that are parabolas, without ever been told what a parabola is (other than the graph of a quadratic function). In the previous chapter, we gave a definition for a parabola, in terms of distances away from a focus and a directrix. Now in this class activity, we asked you to show if you write down an equation for a parabola using this definition, then the equation will in fact represent a quadratic function.

Instead of doing that example again, let's look at the parabola from the class activity in that first section on parabolas. We found the following equation for the parabola with focus (3,6) and directrix $y = 2$:

$$\sqrt{(x-3)^2 + (y-6)^2} = y - 2.$$

Now, we admit this doesn't look like a quadratic function, but it is. We can use the Laws of Algebra to rewrite this equation into the standard quadratic form $f(x) = ax^2 + bx + c$, as follows:

$$\sqrt{(x-3)^2 + (y-6)^2} = y - 2$$

$(y-2)^2 = (x-3)^2 + (y-6)^2$		Definition of the square root, provided $y \geq 2$
$y^2 - 4y + 4 = x^2 - 6x + 9 + y^2 - 12y + 36$		Distributive Law
$y^2 - 4y + 4 = x^2 - 6x + y^2 - 12y + 45$		$9 + 36 = 45$
$-4y + 4 = x^2 - 6x - 12y + 45$		Subtract y^2 from both sides
$8y + 4 = x^2 - 6x + 45$		Add $12y$ to both sides
$8y = x^2 - 6x + 41$		Subtract 4 from both sides
$y = \frac{1}{8}(x^2 - 6x + 41)$		Divide both sides by 12
$y = \frac{1}{8}x^2 - \frac{6}{8}x + \frac{41}{8}$		Distributive Law

So we have re-written the equation for this parabola in the standard form $f(x) = ax^2 + bx + c$, confirming (for this example at least), that the equation for a parabola is indeed a quadratic function.

In the standard form for a quadratic function $f(x) = ax^2 + bx + c$, do the constants a, b and c have any significance? Well, not much. What we know about his parabola from our previous work in Section 21 is that it has a vertex at (3,4), focus at (3,6) and a directrix at $y = 2$. None of these features can be read off immediately from equation $y = \frac{1}{8}x^2 - \frac{6}{8}x + \frac{41}{8}$. However, based on the class activity, we found that if a quadratic function is written in the form $f(x) = a(x - h)^2 + k$, then the vertex of the parabola will be at the point (h, k). That's why $f(x) = a(x - h)^2 + k$ is usually called the vertex form of a quadratic function.

Let's take now rewrite the equation for the parabola in our example into vertex form. Since the vertex form involves $(x - h)^2$ we will use our method of completing the square to do this.

$y = \frac{1}{8}x^2 - \frac{6}{8}x + \frac{41}{8}$		"Standard form"
$8y = x^2 - 6x + 41$		Multiply both sides by 8
$8y - 32 = x^2 - 6x + 9$		Subtract 32 from both sides. We want just 9 to complete the square on the right.
$8y - 32 = (x - 3)^2$		Distributive Law. (Factoring)
$8y = (x - 3)^2 + 32$		Add 32 to both sides.
$y = \frac{1}{8}(x - 3)^2 + 4$		Divide both sides by 8. "Vertex Form"

So there we go. Notice how everything we "did" to the equation to get the right side as a square (namely, divide by 8, and subtract 32) we then later had to "undo" in reverse order (namely add 32 and multiply by 8) to get the equation back to being solved for y. This is yet another example of "doing/undoing" in algebra.

So we've now written the equation for our parabola in "vertex form":

$$f(x) = \tfrac{1}{8}(x - 3)^2 + 4,$$

and we can note that the location of the vertex (3, 4) can be read directly from the function formula. However, the location of the focus and directrix is no longer apparent. Each form for the function reveals different information. That's why it's valuable to be able to use the Laws of Algebra to rewrite function equations into different forms.

To sum up this example, we have written the equation for this one parabola in three different ways, as we will show in the following table:

Equation	Name
$\sqrt{(x-3)^2 + (y-6)^2} = y - 2.$	Focus-Directrix Form
$y = \frac{1}{8}(x-3)^2 + 4$	Vertex Form
$y = \frac{1}{8}x^2 - \frac{6}{8}x + \frac{41}{8}$	Standard Form

As we noted earlier, that "Standard Form" $f(x) = ax^2 + bx + c$, for the quadratic function doesn't say much about its graph. The coefficients a, b and c don't readily tell you the location of the focus, directrix, vertex, or roots of the parabola. So what if we are given a quadratic function in standard form and want to know these things? One option is to use the Laws of Algebra to rewrite the function we have into one into one of these other forms. But if this situation comes up a lot, it would be worthwhile to come up with formulas for the directrix, focus, vertex and roots in terms of the coefficients a, b and c.

In most applications, the vertex and roots are the most useful things to know. In the problem set, we'll ask you to put $f(x) = ax^2 + bx + c$ into vertex form to find a formula for the vertex in terms of a, b and c. Right now, we'll show you how we can find the roots of the function $f(x) = ax^2 + bx + c$ by completing the square.

$$ax^2 + bx + c = 0 \qquad \text{The roots are where } f(x) = 0$$

$$ax^2 + bx = {}^-c \qquad \text{Subtract } c \text{ from both sides}$$

$$x^2 + \frac{b}{a}x = \frac{{}^-c}{a} \qquad \text{Divide both sides by } a$$

$$x^2 + \frac{b}{a}x + \left(\frac{b}{2a}\right)^2 = \frac{{}^-c}{a} + \left(\frac{b}{2a}\right)^2 \qquad \text{Add } \left(\frac{b}{2a}\right)^2 \text{ to both sides}$$

$$\left(x + \frac{b}{2a}\right)^2 = \frac{{}^-c}{a} + \left(\frac{b}{2a}\right)^2 \qquad \text{Distributive Law (factoring)}$$

At this point we have completed the square on the left side. But before we continue to solve for x by undoing that square, it will be worthwhile to simplify that right hand side a bit first.

$$\left(x + \frac{b}{2a}\right)^2 = \frac{{}^-c}{a} + \frac{b^2}{4a^2} \qquad \text{Square } \frac{b}{2a}$$

$$\left(x + \frac{b}{2a}\right)^2 = \frac{{}^-4ac}{4a^2} + \frac{b^2}{4a^2} \qquad \text{Get common denominator}$$

$$\left(x + \frac{b}{2a}\right)^2 = \frac{b^2 - 4ac}{4a^2} \qquad \text{Add fractions}$$

Is that right side starting to look familiar? Now we are ready to solve for x.

$$x + \frac{b}{2a} = \pm\sqrt{\frac{b^2-4ac}{4a^2}}$$ By definition of the square root

$$x + \frac{b}{2a} = \pm\frac{\sqrt{b^2-4ac}}{2a}$$ Simplifying the square root.

$$x = \frac{^-b}{2a} \pm \frac{\sqrt{b^2-4ac}}{2a}$$ Subtract $\frac{b}{2a}$ from both sides

$$x = \frac{^-b \pm \sqrt{b^2-4ac}}{2a}$$ Add fractions

So we've shown that the equation $ax^2 + bx + c = 0$ has two solutions, $x = \frac{^-b+\sqrt{b^2-4ac}}{2a}$ and $x = \frac{^-b-\sqrt{b^2-4ac}}{2a}$. In other words, those are the two roots of the function $f(x) = ax^2 + bx + c$. Of course, what we've just done is derived the "quadratic formula" that you've likely memorized long ago. Well, now you know where it comes from, by solving a generic quadratic equation by completing the square.

Here's a summary of the three different algebraic forms in which you might want to write a quadratic function (or an equation for a parabola). The problem set will help you to fill in the missing details.

Common Forms of Equations for a Vertical Parabola with variables x and y.

Form Name	Equation	Interpretation of Parameters
Standard	$y = ax^2 + bx + c$	Sign of a indicates direction of opening
		Roots are at: $x = \frac{^-b+\sqrt{b^2-4ac}}{2a}$ and $x = \frac{^-b-\sqrt{b^2-4ac}}{2a}$.
		Vertex is at:
Vertex	$y = a(x-h)^2 + k$	Sign of a indicates direction of opening
		Vertex is at (h, k)
		Roots are at:
Focus-directrix	See problem set	Focus is (b, c)
		Directrix is $y = d$

Homework Set 23

1. Consider the quadratic function $f(x) = 2(x+3)^2 - 4$
    a) Sketch the graph of this function.
    b) Find the exact location of the vertex and roots of the parabola.
    c) Write the function in the standard quadratic form $f(x) = ax^2 + bx + c$.

2. Consider the quadratic function $g(x) = {}^-2(x-1)^2 + 5$
    a) Sketch the graph of this function.
    b) Find the exact location of the vertex and roots of the parabola.
    c) Write the function in the standard quadratic form $f(x) = ax^2 + bx + c$.

3. Consider the quadratic function $f(x) = 2x^2 + 12x + 19$
    a) Complete the square to write this function into vertex form.
    b) Use the vertex form to find the roots of this function.
    c) Sketch a graph of this function, showing the location of the vertex, roots, and two other points on the parabola.

4. Consider the quadratic function $g(x) = {}^-2x^2 + 5x + \frac{1}{4}$
    a) Complete the square to write this function into vertex form.
    b) Use the vertex form to find the roots of this function.
    c) Sketch a graph of this function, showing the location of the vertex, roots, and two other points on the parabola.

5. Show how to solve the following equation by using three methods (Factoring, Completing the Square, Quadratic Formula).
$$6x^2 = x + 2$$

6. The height of a ball after t seconds, if it is thrown straight up from a height of h_0 feet and initial velocity v_0 feet per second, according to Newton's Law of gravity, will be modeled by the following function:
$$h(t) = -16t^2 + v_0 t + h_0$$

    Imagine you throw a baseball straight up with an initial velocity of 32 feet per second, and you release the ball at time 0 from a height of 6 feet above the ground.

    a. How high up into the air will the ball go?
    b. At what time will the ball reach this maximum height?
    c. At what time will the ball hit the ground?
    d. Sketch a graph of the height of the ball from $t = 0$ until it hits the ground.

7. The data for Colony III from the Class Activity in the previous section was modeled by the quadratic function $P(t) = 12t^2 - 4t + 50$. Complete the square to write this in vertex form. Sketch a graph of the function that shows the vertex, as well as the data from the Class Activity.

8. In a previous section's homework, you graphed and found equations for the following parabolas. Now show how you can write a sequence of equivalent equations to rewrite these in vertex form, and verify that vertex of the parabola shows up correctly in that form.
    a) The parabola with directrix $y = {}^-6$ and focus at $(4,0)$.
    b) The parabola with a vertex at $(0,6)$ and a focus at $(0,2)$.

9. The vertex form of a quadratic function makes it easy to find its vertex and its roots. We already noted in the Read and Study that the vertex for $f(x) = a(x-h)^2 + k$ will be at (h, k). Now find the roots for $f(x) = a(x-h)^2 + k$ and add this information to the summary table in the Read and Study.

10. Show a sequence of equivalent equations to rewrite $f(x) = ax^2 + bx + c$ into vertex form. (Hint: Complete the square on $y = ax^2 + bx + c$. Use the derivation of the quadratic formula in the Read and Study as a reference.)

11. Use the result of the previous problem to find a formula for the vertex of the parabola given by the function $f(x) = ax^2 + bx + c$, in terms of a, b and c, and add this information to the table in the Read and Study.

12. Write an equation for the parabola with focus (b, c) and directrix $y = d$, then add this to the table in the Read and Study.

Class Activity 24a: Transformational Thinking

Starting with the graph of $f(x) = x^2$, sketch the graphs of the given transformations of f. Use a different color for each graph. Describe how each graph is related to the previous graph in the sequence, and how the location of the vertex changes with each transformation.

$y = x^2$ (black)

x	y
-2	
-1	
0	
1	
2	

←vertex

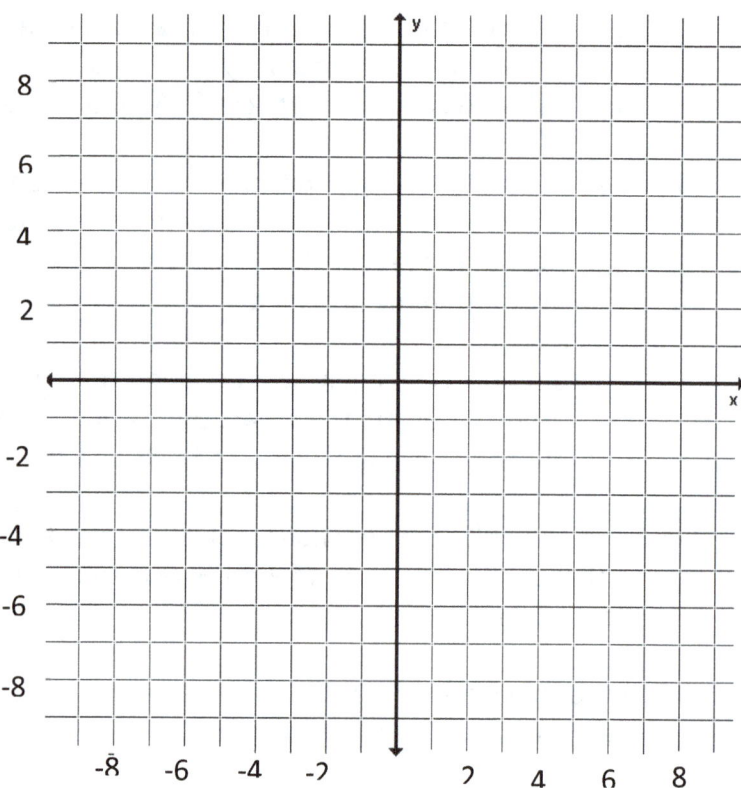

$y = (x + 4)^2$ (red)

x	y

$y = {}^-2(x + 4)^2$ (green)

x	y

$y = {}^-2(x + 4)^2 - 3$ (blue)

x	y

Make conjectures about what effect the parameters a, h, and k have in determining how the graph of $y = a(x - h)^2 + k$ is the result of transforming the graph of $f(x) = x^2$.

Class Activity 24b: On the Move

The complete graph of a function $y = f(x)$ is shown here:

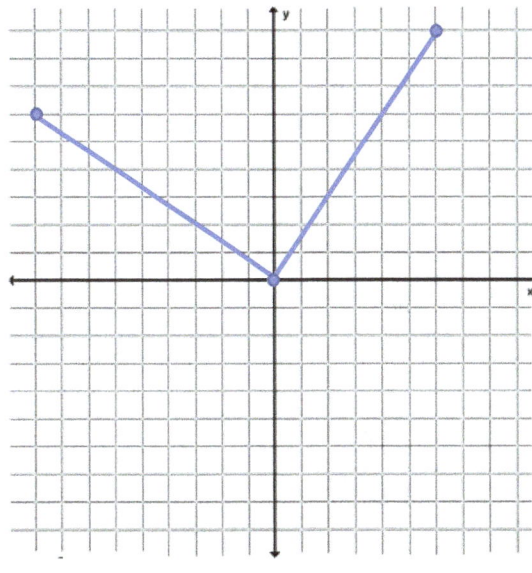

For each function below:
- Make a function machine diagram (one of the operations will be the function f).
- Use your function machine diagram to help generate a table of inputs and outputs for the new function
- Sketch the graph of the new function.
- Describe how the graph of the new function is related to the graph of $f(x)$. Be specific.

$a(x) = f(2x)$

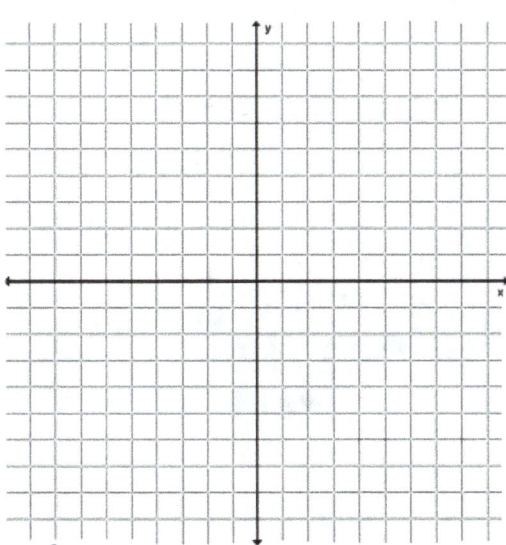

177

$b(x) = f(2x + 3)$

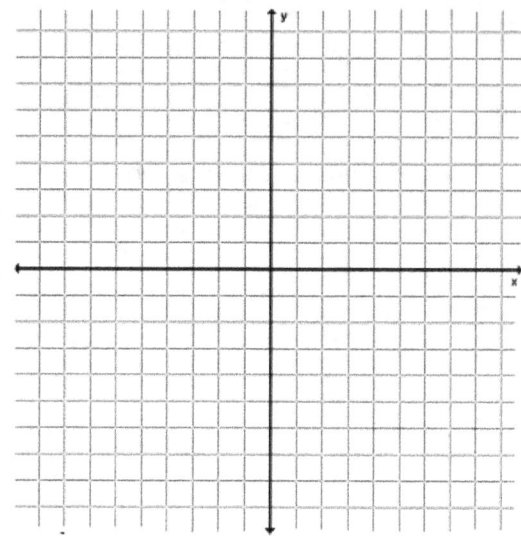

$c(x) = {}^-f(2x + 3)$

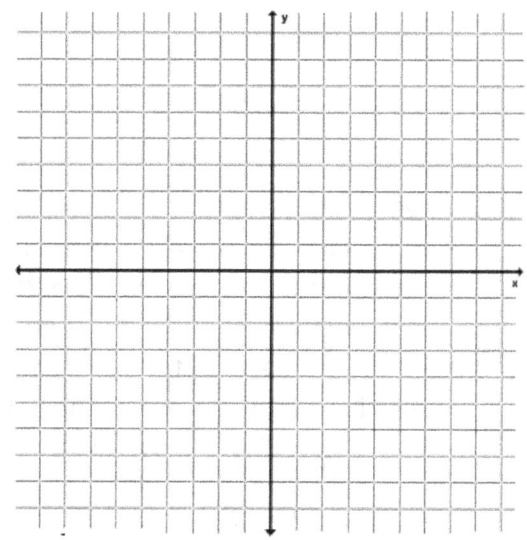

$d(x) = {}^-f(2x + 3) + 4$

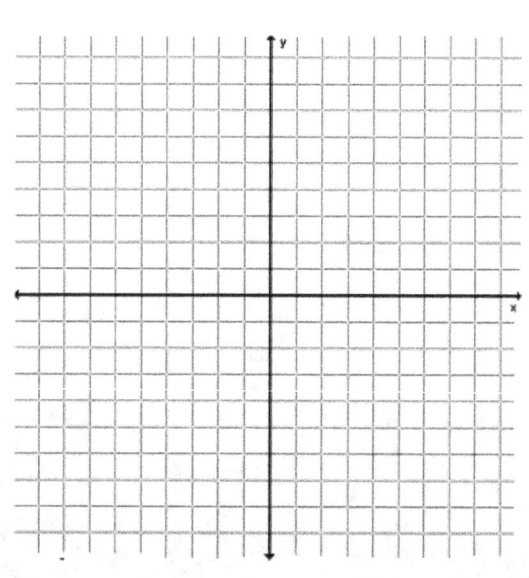

Read and Study 24: Transformations of Functions

All things are difficult before they are easy.

John Norley

In the Class Activity, we saw how the graph of a function can be transformed by sliding it up or down vertically, sliding it left or right horizontal, and by stretching or contracting it. Another transformation that we will see is a reflection. Each of these transformations can be accomplished by applying some other function that adds, subtracts, multiplies or divides by a number, either before the input goes into f, or after the output comes out of f. When an input has first one function done to it, then that output goes into another function, that's called a **composition** of functions.

Here's a machine diagram for the composition $g[f(x)]$. First x is put into the function f, then the resulting output $f(x)$ is put into the function g.

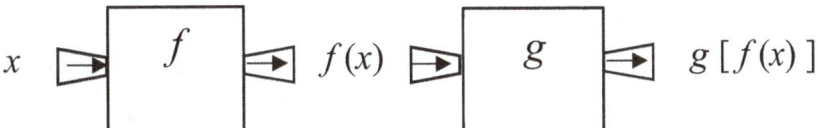

In contrast, here's a machine diagram for the composition $f[g(x)]$:

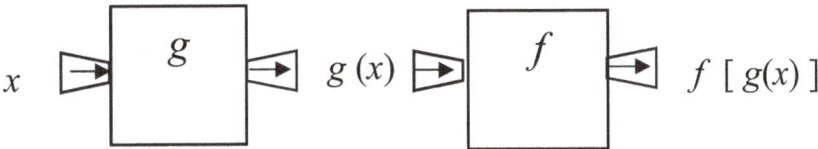

In this read and study, we'll look at the transformations in the first Class Activity again, and view those transformations of $f(x) = x^2$ as composing f with some other function g. The first transformation of $y = x^2$ we looked at was $y = (x + 3)^2$. Here the number 3 is added to x first, before the result goes into the squaring function f. If g is the "add three "function, and f is the "square" function, the composition we want is $f[g(x)]$, and the machine diagram for this transformation would be:

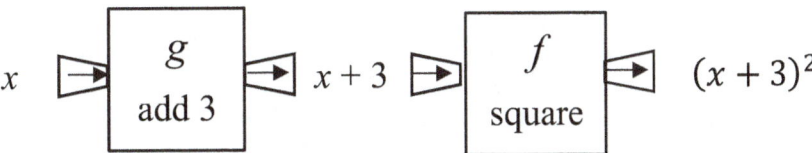

As we saw in the Class Activity, since this transformation adds 3 to the x value *before* it goes into f, this transformation has the effect of shifting the graph of $f(x) = x^2$ to the left by three units.

The order of the composition here is crucial. If instead we had put x first into the squaring function f, then put the result into the "add 3" function g, then the composition would be $g[f(x)]$ which looks like this:

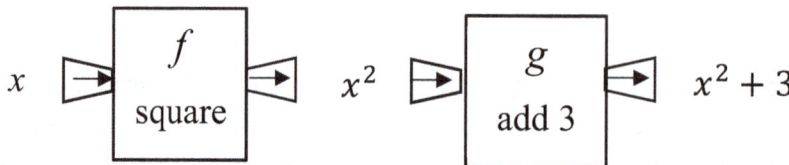

The graph of the resulting function $y = x^2 + 3$, since it adds three to the y value, has the effect of shifting the graph of $f(x) = x^2$ up by three units.

Fill in the tables below to compare the effect of either adding the three first to the x value before squaring, or adding the three at the end to the y value (after squaring).

x	$x + 3$	$(x + 3)^2$	x	x^2	$x^2 + 3$
-3			-3		
-2			-2		
-1			-1		
0			0		
1			1		
2			2		
3			3		

Shown below are graphs of $y = x^2$, $y = (x + 3)^2$ and $y = x^2 + 3$. Use the data in the tables to label which is which, and describe how the graph of $f(x) = x^2$ is shifted by either adding 3 before or after x is put into the function.

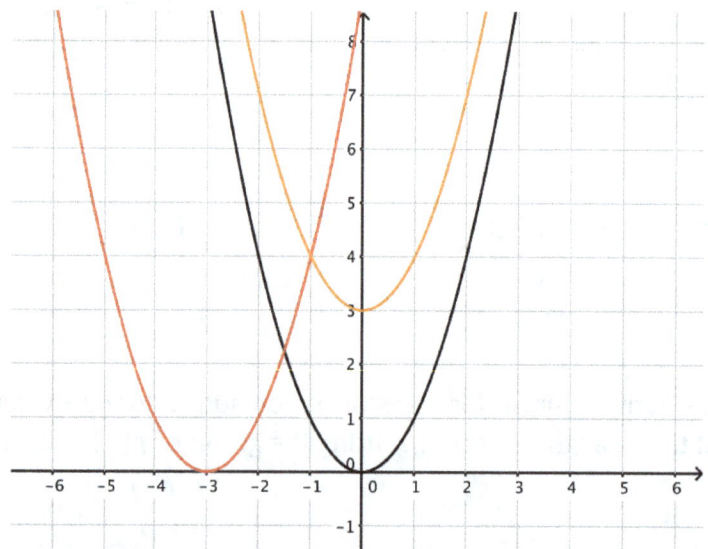

In the second Class Activity that follows this Read and Study, we will have you look further at the effect that **multiplying** by a number either before or after applying the function $f(x)$ has on the graph. Pay attention to the effect on the graph multiplying by a number that is greater than 1 versus a number that is smaller than 1, and of course, whether you are multiplying before or after function f is applied. Summarize what you find in the table at the end of this section.

We'll finish the Read and Study by looking at a special case of composing a function with a multiplication: multiplying by $^-1$.

Let's first consider the graph of $y = {}^-1 \cdot x^2$. Here x first goes into the squaring function, then the number $^-1$ is multiplied afterwards to x^2. If g is the "multiply by $^-1$ "function, and f is the "square" function, the composition we want is $g[f(x)]$, and the machine diagram for this transformation would be

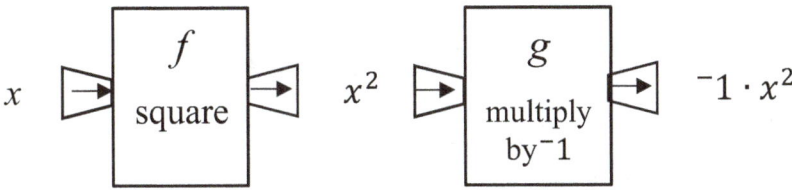

In the table below, we show some input values for x, the intermediate outputs x^2 and final outputs $^-1 \cdot x^2$. The graph of $y = {}^-1 \cdot x^2$ is then shown in red. Comparing this graph with the graph of $y = x^2$, we see that the transformation is to reflect the graph across the x-axis.

x	x^2	$^-1 \cdot x^2$
$^-3$	9	$^-9$
$^-2$	4	$^-4$
$^-1$	1	$^-1$
0	0	0
1	1	$^-1$
2	4	$^-4$
3	9	$^-9$

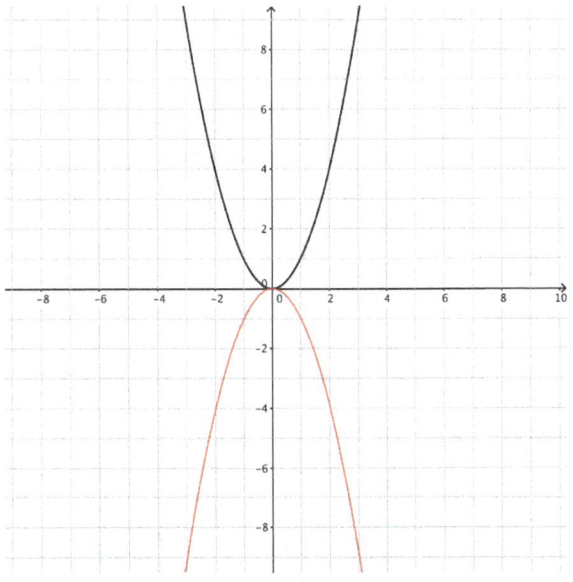

Now let's consider the graph of $y = (^-1 \cdot x)^2$. Now the number $^-1$ is multiplied to x first, before the result goes into the squaring function f. If g is the "multiply by $^-1$ "function, and f is the "square" function, the composition we want is $f[g(x)]$, and the machine diagram for this transformation would be

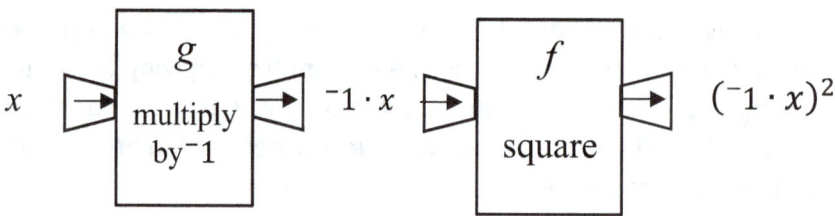

In the table below, we show some input values for x, the intermediate outputs $^-1 \cdot x$ and final outputs $(^-1 \cdot x)^2$. Since these are the same outputs we'd get for $y = x^2$, we see that the graph of $y = (^-1 \cdot x)^2$ will look exactly like the graph of $y = x^2$. Actually, there *is* a transformation that took place, namely to reflect the graph across the y-axis. But since the graph of $y = x^2$ is symmetric with respect of the y-axis, the graph ends up reflecting back onto itself. To try to illustrate this, we will graph each "half" of the parabola in a different color and show where each half reflects to. If the graph of $y = x^2$ is shown below left, then the graph of $y = (^-1 \cdot x)^2$ is shown below right.

x	$^-1 \cdot x$	$(^-1 \cdot x)^2$
$^-3$	3	9
$^-2$	2	4
$^-1$	1	1
0	0	0
1	$^-1$	1
2	$^-2$	4
3	$^-3$	9

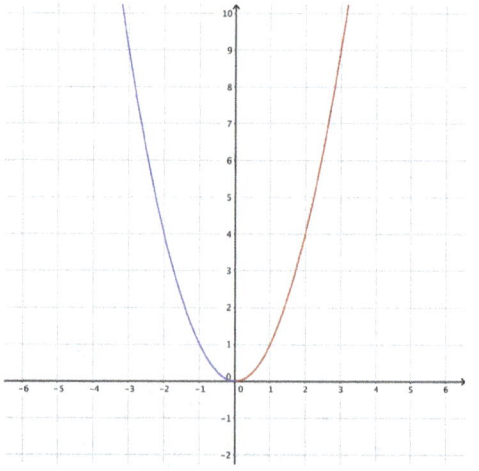
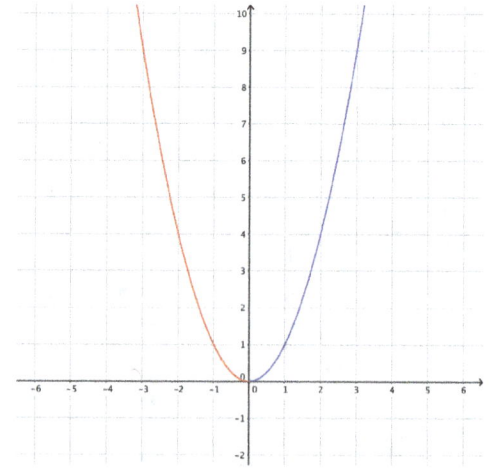

Notice that with multiplication again the order of the composition is crucial. Multiplying x by negative 1 before it's put into the function f, results in reflecting the graph across the y axis, while multiplying by negative 1 after the output comes out of f results in reflecting the graph across the x axis.

Homework Set 24

1. The function $y = f(x)$ is defined by the following table. Note, the domain of this function is **only** the numbers in the set {0,1,2,3,4,5,6}. Sketch the graph of the function $f(x)$. Then make the table and sketch the graph of the following functions:

   a. $a(x) = f(x) + 3$

   b. $b(x) = f(x + 3)$

   c. $c(x) = f(3x)$

   d. $d(x) = 3f(x)$

   e. $g(x) = f\left(\frac{1}{3}x\right)$

   f. $h(x) = \frac{1}{3}f(x)$

x	$f(x)$
0	-4
1	0
2	4
3	8
4	7
5	6
6	5

2. Sketch the graph of the function $f(x) = x^2$ on the grid below. Then, using a different color pen, sketch the graph of function $g(x) = \frac{1}{2}(x - 1)^2 - 5$ on the same grid.

   a. Locate the **vertex** of the graph of the function $g(x) = \frac{1}{2}(x - 1)^2 - 5$

   b. Use the formula $g(x) = \frac{1}{2}(x - 1)^2 - 5$ to find the exact roots of this function.

   c. Locate of the **roots** of the function $g(x) = \frac{1}{2}(x - 1)^2 - 5$ on the graph.

   d. Describe how the graph of $f(x) = x^2$ can be transformed to create the graph of $g(x) = \frac{1}{2}(x - 1)^2 - 5$.

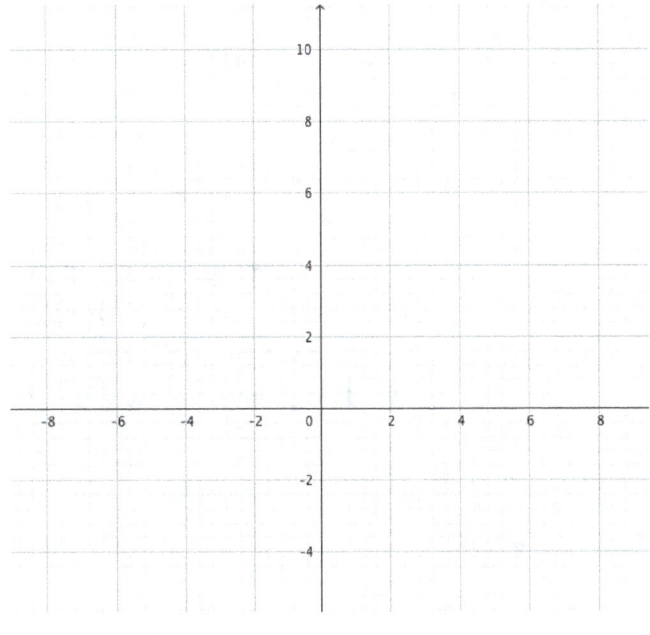

183

3. If $g(x) = 3x^2 - 2x$, find a formula for:
   a. $g(x^2)$
   b. $(g(x))^2$
   c. $g(x-4)$
   d. $g(x) - 4$

4. Find a possible formula for $g(x)$ if:
   a. $g(x^3) = 2x^6 + 1$.
   b. $g(x+1) = x^2 + 2x + 1$

5. For each function below, determine whether $f(a+b) = f(a) + f(b)$ is valid in general (true for any a and b in the domain of f). If it is valid, prove it using the properties of arithmetic. If it is not valid, prove it by giving an example of an a and b so that $f(a+b) \neq f(a) + f(b)$.

   a. $f(x) = x^2$
   b. $f(x) = \sqrt{x}$
   c. $f(x) = \frac{1}{x}$
   d. $f(x) = 3x + 5$

6. For each function below, determine whether $f(cx) = c \cdot f(x)$ is valid in general (true for any a and b in the domain of f). If it is valid, prove it using the properties of arithmetic. If it is not valid, prove it by giving an example of a c and x so that $f(cx) \neq c \cdot f(x)$.

   a. $f(x) = x^2$
   b. $f(x) = \sqrt{x}$
   c. $f(x) = \frac{1}{x}$
   d. $f(x) = 3x + 5$

7. In the Read and Study, we said the order of the composition is crucial. Usually $g(f(x))$ will not be the same as $f(g(x))$.
   a. Find an example of two functions $f(x)$ and $g(x)$ such that $g(f(x)) \neq f(g(x))$.
   b. Find an example of two functions $f(x)$ and $g(x)$ such that $g(f(x)) = f(g(x))$.

Class Activity 25a: Popcorn Boxes

We are going to make open-top boxes by cutting four squares off the corners of an 8.5 by 11 inch piece of paper and folding up the sides. The plan is to find the box with maximum volume. Start by making boxes where the squares you cut away have side lengths 1, 1.5, 2, 2.5 and 3 inches.

1. Make predictions. Just by looking at them, guess the order of the boxes in terms of increasing volume.
2. Now do the calculations; that is, find the volume of each of the 5 boxes. Which one has the largest volume. How good were your predictions?
3. Is there a box you haven't made yet that has the largest possible volume? If so, find its dimensions. If not, why not?
4. Find an algebraic model for the volume of the box as a function of the "cut away side length"
5. Sketch a graph of the function you found in part 4. Find and label the roots. Use the graph to find the side length that gives the largest volume.

Class Activity 25b: Fill 'er Up!

Suppose water is poured into each container below at a constant rate. For each, sketch a graph of the height of the water in the container over time.

For each function, decide whether it might be modeled by one of the function forms we've discussed so far: linear, quadratic, or exponential. Explain your reasoning.

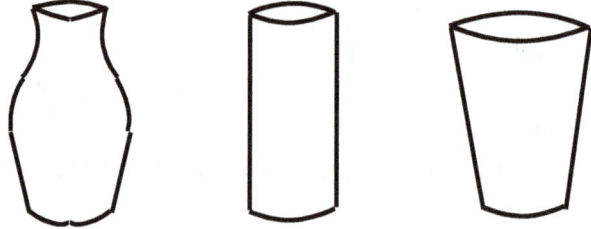

Read and Study 25: The Fundamental Theorem of Algebra

Mathematics is the art of giving the same name to different things.

Poincare

Linear function and quadratic functions belong to a specific class of functions called polynomials. A **polynomial** is a function that can be put in the form:

$$f(x) = a_n x^n + a_{n-1} x^{n-1} + a_{n-2} x^{n-2} + \cdots + a_2 x^2 + a_1 x^1 + a_0,$$

where n is some natural number (called the **degree** of the polynomial) and all the 'a's are real numbers (called the **coefficients**). The subscripts are just part of their names to label them and to show that they may all be different numbers. We require that a_n not be zero. *Why?*

For example, $g(x) = 2x + 8x^3$ is a polynomial of degree 3. *Decide whether each of the following fits the definition of a polynomial, and if so, state its degree:*

1) $h(x) = x^{\frac{1}{2}} + 3x^2 + 5$

2) $j(x) = 0.34x^9 - 546x^3 + 5$

3) $k(x) = 9(5 - x)(x + 7)(x - 3)^2$

4) $m(x) = \frac{x^2+1}{x-2}$

Okay, so only the second and the third above qualify as polynomials. The second has degree 9 and the third has degree 4 (*multiply it out and see*). The first function has an exponent that is not a natural number so it cannot be a polynomial. The last function is a quotient of two polynomials, this kind of function is called a "rational function" and we'll look at those in the next section.

Polynomials have lots of great properties. A lot of them stem from the **Fundamental Theorem of Algebra**. It guarantees that every n-degree polynomial

$$p(x) = a_n x^n + a_{n-1} x^{n-1} + a_{n-2} x^{n-2} + \cdots + a_2 x^2 + a_1 x^1 + a_0,$$

has exactly n roots (counting double roots as two, triples roots as three, etc.) if we allow complex numbers as roots. Furthermore, if we label the roots $r_1, r_2, \ldots, r_n$, then the polynomial can be written as a product of linear factors as follows:

$$p(x) = a_n(x - r_1)(x - r_2)(x - r_3) \cdots (x - r_n).$$

In our discussion of polynomials, it's important to make distinctions about the terminology we are using:

- We say that $x = 2$ is a **solution** to the equation $x^2 - 4 = 0$.

- We say $(x - 2)$ is a **factor** of the polynomial $f(x) = x^2 - 4$.

- We say that $x = 2$ is a **root** of the polynomial $f(x) = x^2 - 4$.

How many roots can polynomial functions have? The Fundamental Theorem of Algebra says that every polynomial of degree n has exactly n roots (counting multiple roots) if we allow complex numbers as roots. So this means that a degree-seven polynomial can have at most seven real roots, and that it has exactly seven roots if count complex numbers. In 1799 a mathematician named Gauss proved this theorem; he was twenty-two years old. Today there are over 100 different proofs that people have made of this fundamental theorem.

Polynomials in completely factored form, like number three above, are fun to graph. We'll talk you through that one so you can see how we think about it.

Let $k(x) = 9(5 - x)(x + 7)(x - 3)^2$. The roots of this polynomial are $x = 5$, $x = {}^-7$ (those are both single roots) and $x = 3$ (a double root). So we can begin by plotting these roots on the graph. This is where the function has a value of zero.

Now we can check whether the value for the function will be positive or negative between each of these roots, by checking test cases in those intervals. For example, at x = 0, we can easily see that the function value will be $k(0) = 9 \cdot 5 \cdot 7 \cdot ({}^-3)^2$, which we know is a positive number. So now we know the polynomial will be positive in that entire interval between the roots of $x = {}^-7$ and $x = 3$. Why? Well, another great property of polynomials is that they are **continuous.** Informally this means that the graph of a polynomial will not have any holes, breaks or vertical asymptotes. Since polynomials are continuous, the graphs can't "jump" over the x-axis. Whenever they change from positive to negative, they have to do so by crossing the x-axis at a root.

Similarly, we can test an x-value between 3 and 5 (say, $x = 4$), to see that the function is also positive on that interval (since $k(4) = 9 \cdot 1 \cdot 1 \cdot (1)^2$, which is positive. We can then test an x-value beyond the largest root at 5 (say at $x = 10$, where we get $k(10) = 9 \cdot ({}^-5) \cdot 17 \cdot (7)^2$, which is a negative number. So we know the function will always be negative for all x greater than 5. How do we know it has to stay negative for all x greater than five? Well, by the continuity of polynomials, if it were to become positive again for some x greater than 5, there would have to be a point where the graph crosses the x-axis again, making another root. But we know all the roots and know there aren't any more roots greater than 5. So the function has to stay negative. Similarly, testing an x-value less than ${}^-7$, say at $x = {}^-10$, we get $k({}^-10) = 9 \cdot (15) \cdot ({}^-3) \cdot ({}^-13)^2$ which is a negative number, so the function stays negative for all x less than ${}^-7$.

The **end behavior** of a function is a description of what happens to the y values of the function as the x values tend to infinity and tend to negative infinity. The end behavior of **polynomial** functions is that they always tend to positive or negative infinity as x does. This is because if we put in larger and larger values for x, each factor will also get larger and larger (either more positive or more negative) and the product of all these factors will get larger and larger (more positive or more negative

In summary, we have figured out that $k(x) = 9(5-x)(x+7)(x-3)^2$ will tend to negative infinity as x tends to negative infinity, that it will be zero at $x = {}^-7$, positive between $x = {}^-7$ and $x = 3$, zero at $x = 3$, positive between $x = 3$ and $x = 5$, zero at $x = 5$, and tend to negative infinity as x tends to infinity, all just by looking at the factored form of the polynomial and testing out values. Using this information we can then sketch a graph of the function by making a continuous curve through the roots that is positive (above the x-axis) and negative (below the x-axis) on the proper intervals between the roots, like this:

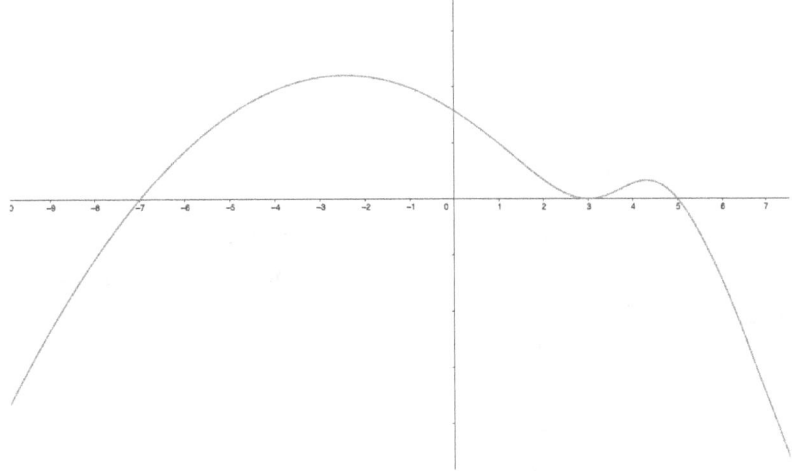

Have a look at the behavior of the graph near the roots. The factored form of the polynomial $k(x) = 9(5-x)(x+7)(x-3)^2$ also tells you *how* the graph of the function hits those roots. At $x = 5$ and $x = {}^-7$, these are single roots, and the graph goes through the x-axis (like a straight line would); at $x = 3$, which is a double root, the graph will touch then bounce off the x-axis (like the vertex of a parabola would).

A polynomial that is in completely factored form is pretty special. In general, it is tough to factor a polynomial completely. For one thing, not all polynomials can be factored unless you allow complex numbers. Take the function $f(x) = x^2 + 4$. It is a polynomial of degree 2 and so by the Fundamental Theorem of Algebra, it has 2 roots. But the graph never crosses the x-axis. *Sketch its graph it and see.* If you solve $x^2 + 4 = 0$ to find its roots, you find the solutions are $2i$ and ^-2i. So we say this function has two complex roots.

For another thing, even if a polynomial had only real roots, we do not have too many tools for factoring them. Degree 1, 2, 3 and 4 polynomials are the exceptions. There are formulas for factoring these. (In the case of degree 2 polynomials, that formula is called the quadratic formula, which can be derived by completing the square.) Similarly, one can derive a formula that will work to factor any degree 3 polynomial, and a formula to factor any degree 4 polynomial. (But get this: mathematicians have *proved* that no such formulas exist for general polynomials with degrees higher than four. It isn't that we haven't found them yet. They cannot exist. That's pretty amazing that we can prove that!)

It turns out that solving polynomial equations of degree three and higher, with or without a formula, is difficult (and even impossible sometimes). So for the most part, people solve these equations numerically or graphically using technology. Solving any equation graphically has its own difficulties, however. The most important thing when solving an equation graphically is to be sure that you are looking at the whole picture. In other words, the key is making sure you have found *all* of the solutions.

Homework Set 25

1) Based on the definition, is the following function a polynomial? Why or why not? If so, what is its degree? What is the coefficient of the x^2 term?

$$f(x) = 3x(x-4)^2(x+2)$$

2) Sketch, *without a calculator*, graphs of the following functions of x. Don't worry about the scale on the y axis, but show the location of all roots and the general behavior of the function between the roots.
   a. $y = 4(x-6)(x+3)(x-8)$
   b. $y = 3x(x-4)^2(x+2)$
   c. $y = 2x^3(x-5)(x+4)$
   d. $y = \frac{1}{2}(x+3)^2 - 6$

3) The graph of a function $y = f(x)$ is sketched below. (The y axis is not shown to scale). Write an appropriate algebraic formula for the function.

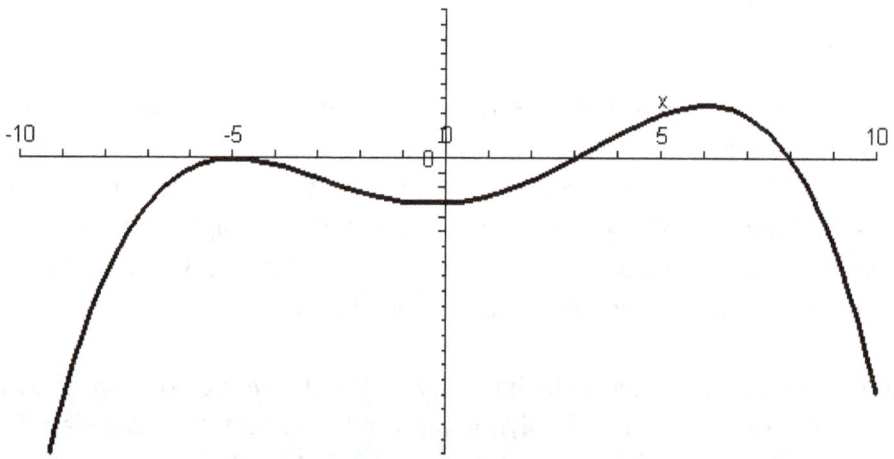

190

3) Consider the cubic polynomial function $f(x) = x^3 + 6x^2 + 10x + 4$.
   a) Show that $x = {}^-2$ is one root of this function.
   b) Since $x = {}^-2$ is one root, you know that $(x + 2)$ is one of the factors. Use this fact to completely factor $f(x)$.
   c) Use the factored form to list the exact values of all of the roots of this function.
   d) Sketch the graph of this polynomial, showing the roots and end behavior.

4) Consider the equation $x^3 - 2x = 5x^2 - 6$. One solution to this equation is a small integer. Find it by trial and error. Then use the method from the previous problem to find all solutions to the equation.

5) For each polynomial function given, write the polynomial as a product of linear factors, and list all the complex roots of the function. Sketch the graph of the function.
   a) $f(x) = x^3 + 3x^2 + 12x$
   b) $g(x) = (x + 4)^2 - 7$
   c) $h(x) = x^4 + 2x^2 - 8$
   d) $p(x) = x^3 - 7x + 6$
   e) $q(x) = x^4 - 81$

6) Build Your Own Polynomial. Write a function formula and sketch the graph of your function that satisfies the given condition.
   a. Linear polynomial with exactly one real root $x = 6$.
   b. Quadratic polynomial with exactly one real root $x = 6$.
   c. Cubic polynomial with exactly one real root $x = 6$.
   d. Linear Polynomial with no real roots
   e. Quadratic Polynomial with no real roots

7) More Build Your Own. Write a function formula or equation that satisfies the following conditions:
   a. Polynomial written in the form $y = ax^2 + bx + c$ with roots $x = \frac{5+\sqrt{3}}{2}$ and $x = \frac{5-\sqrt{3}}{2}$.
   b. Equation with exactly 3 solutions: $x = 0$, $x = 3$, $x = 6$.
   c. Polynomial with roots at exactly $x = {}^-6$, $x = \sqrt{2}$, $x = {}^-\sqrt{2}$, $x = \frac{2}{3}$
   d. Equation with only real coefficients that has solutions $x = 7$, $x = 2i$, $x = {}^-2i$.
   e. Degree 5 polynomial with roots only at 1 and 0.

8) Decide whether each of the following is True or False. In each case, justify your answer.
   a. All third degree polynomials have at least one real root.
   b. All fourth degree polynomials have at least one real root.
   c. If $(x - 3)$ is a factor of a polynomial, then $x = 3$ is a root.
   d. Complex roots of polynomials always come in conjugate pairs.

9) Summarize how the end behavior of a polynomial function depends on the degree of the polynomial.

Chapter Five

Inverse Functions and Non-Polynomial Forms

Class Activity 26: Let's Be Rational About This

A rational function is a fraction of two polynomials. In this activity you will analyze some examples of rational functions and generalize their properties.

1. $f(x) = \frac{2}{x+5}$.

This function does not have any roots. (Why not? How can you tell?)

This function has a discontinuity when $x = {}^-5$. (Why? How can you tell?)

Fill in the table to see what happens as x approaches $^-5$, and as x approaches positive and negative infinity.

x	$f(x)$
-4	
-4.5	
-4.9	
-4.99	
-5	
-5.01	
-5.1	
-5.5	
-6	

x	$f(x)$
-1000	
-100	
-10	
10	
100	
1000	

Use this information to sketch a graph of the function $f(x) = \frac{2}{x+5}$ showing its salient features.

2. Now do a similar analysis on $f(x) = \dfrac{2x-3}{x+5}$ to identity its roots, discontinuities, behavior at the discontinuities and end behavior. Then sketch its graph.

3. Now do a similar analysis on $f(x) = \dfrac{x^2-4}{x+5}$ to identity its roots, discontinuities, behavior at the discontinuities and end behavior. Then sketch its graph.

Compare the features of the graphs of the three functions in this activity. Make some conjectures about the roots and discontinuities and end behavior of rational functions.

Read and Study 26: Rational Functions

Somewhere, something incredible is waiting to be known.

Carl Sagan

A **rational function** is a function that is the ratio (fraction) of two polynomials. For example, $f(x) = \frac{3x^2-21x+18}{x^2-2x-8}$ and $g(x) = \frac{x^5}{x^2+1}$ are examples of rational functions. So is $h(x) = \frac{1}{x}$. Compare this definition for rational *number* being defined as a ratio (fraction) of two integers.

The behavior of a rational function is determined by the behavior of the polynomials that make up the numerator and denominator. For example, a rational function will be zero whenever the numerator polynomial is zero. So the roots of rational function are the roots of its numerator polynomial. And rational function will be undefined whenever the denominator polynomial is zero. That means that any values for x that is a root of the denominator polynomial will not be in the domain of the rational function.

Recall that a nice property of polynomial functions is that they are continuous, meaning that their graphs contain no holes, jumps, or vertical asymptotes. Well, rational functions are also therefore continuous, except at those values x where the denominator polynomial is zero. These points are called "**discontinuities**" for the rational function. Will the function $h(x) = \frac{1}{x}$ have a discontinuity? If so, where? Will the function $g(x) = \frac{x^5}{x^2+1}$ have a discontinuity? If so, where?

As an example, let's consider the function $f(x) = \frac{3x^2-21x+18}{x^2-2x-8}$. To find the roots and discontinuities, we want to find where the numerator and denominator polynomials are zero. Writing these polynomials in factored form will make these easy to find.

$$f(x) = \frac{3x^2 - 21x + 18}{x^2 - 2x - 8}$$

$$= \frac{3(x-1)(x-6)}{(x-4)(x+2)}$$

So we can see that $f(x)$ will have roots at $x = 1$ and at $x = 6$, since that's where the numerator will be zero. Also, we see that $f(x)$ will be undefined when $x = 4$ or $x = -2$, since those values make the polynomial $x^2 - 2x - 8$ in the denominator zero. So $x = 4$ and $x = -2$ are not in the domain of this function, and we say the function will have a discontinuity at each of these x values.

Let's investigate what the behavior of the function is near each discontinuity at $x = 4$ and $x = -2$. We will make a table of values as x gets close to each of these discontinuities. Notice how as x approaches $x = 4$, the function value grows without bound one either side (towards negative infinity on one side and positive infinity on the other). A similar thing happens at $x = {}^-2$. We say that the function has **vertical asymptotes** at $x = 4$ and at $x = {}^-2$.

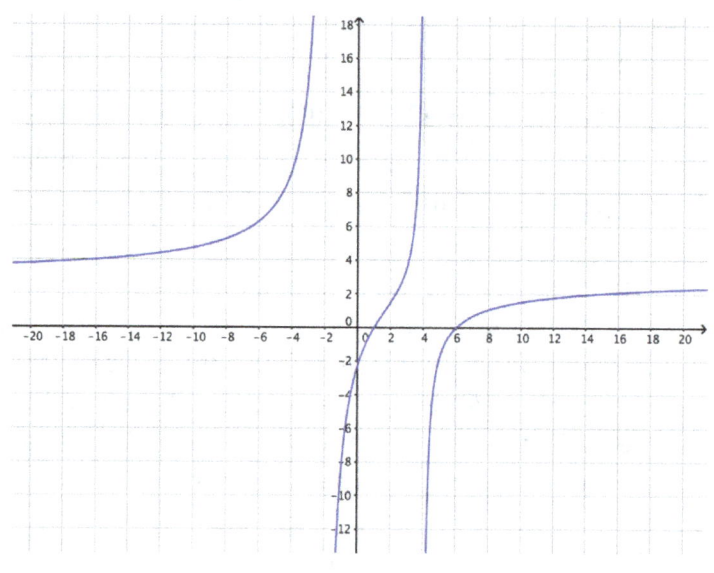

x	$f(x)$
3	3.6
3.5	6.8
3.9	31.0
3.99	301.0
4	undefined
4.01	-299.0
4.1	-29.0
4.5	-4.9
5	-1.7

x	$f(x)$
-3	15.4
-2.5	27.5
-2.01	1203.5
-2.001	12003.5
-2	undefined
-1.99	-1196.5
-1.9	-116.5
-1.5	-20.5
-1	-8.4

The **end behavior** of a function is a description of what happens to the y values of the function as the x values tend to infinity and tend to negative infinity. As we saw in the previous section, the end behavior of **polynomial** functions is that they always tend to positive or negative infinity as x does. For **rational** functions, however, another type of end behavior is possible: the y values can continue to get closer and closer to some finite value. This is called having a **horizontal asymptote**.

Let's investigate the end behavior of our example function $f(x) = \frac{3x^2 - 21x + 18}{x^2 - 2x - 8}$ by inputting some larger and larger values for x:

x	$f(x)$
100	2.851103
1,000	2.985012
10,000	2.998500
100,000	2.999850

x	$f(x)$
-100	3.151295
-1,000	3.015012
-10,000	3.001500
-100,000	3.000150

We can see that as x grows large in either the positive or negative direction, that the function gets closer and closer to the value of 3, so we say the function has a horizontal asymptote at l$y = 3$. n the Class Activity, you hopefully noticed that the end behavior of a polynomial will depend the degrees of the polynomials in the numerator and denominator. We will ask you to summarize this relationship in the problem set.

Homework Set 26

1) Shown below are four graphs, corresponding to the following four functions. Without doing any calculations or plotting points, try to match up each function with its graph, by paying attention to what the roots, discontinuities, and end behavior of the function should be.

$$a(x) = \frac{x}{x^2+5} \qquad\qquad b(x) = \frac{x^2}{x^2+5}.$$

$$c(x) = \frac{x}{(x+5)^2} \qquad\qquad d(x) = \frac{x^2}{(x+5)^2}.$$

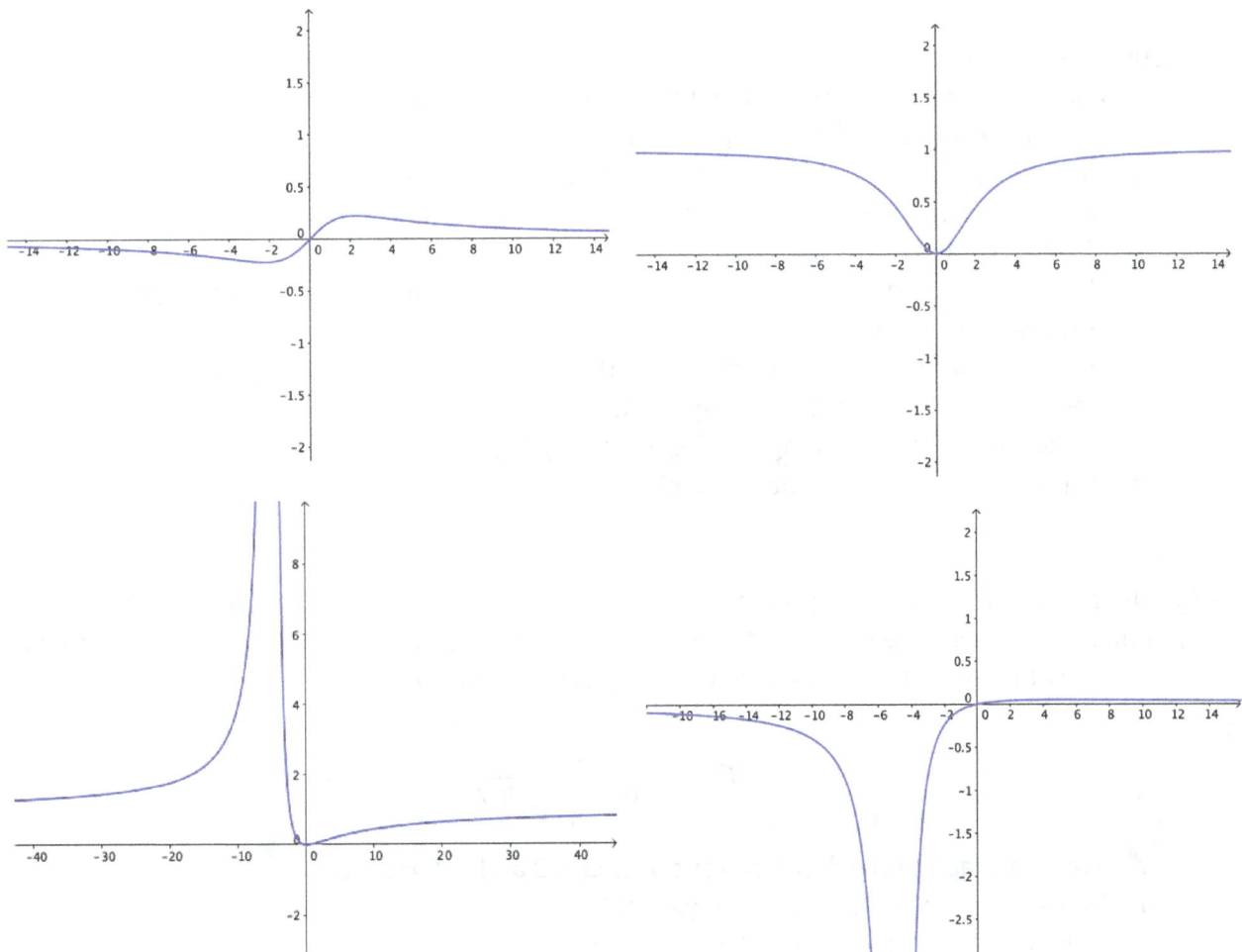

197

2) Find all of the roots and discontinuities for the function. Investigate the behavior of the function between the roots and at the discontinuities, as well as the end behavior of the function. Use this information to sketch a graph that shows all the salient features of the function. (Don't worry about the scale on the y- axis). Also identify any horizontal asymptotes.

   a. $f(x) = \frac{2x^2+20x}{x^2-10x+16}$

   b. $g(x) = \frac{x^3-9x}{x^2+1}$

   c. $h(x) = \frac{1}{x}$

   d. $k(x) = \frac{2x+3}{x^2-5}$

3) Suppose you have a distance of 25 miles to cover. How long this will take depends on the speed at which you travel. Let $f(x)$ be the time in minutes it will take to 25 miles if you travel at a constant rate of x miles per hour.
   a. Make a table of values for $f(x)$ that includes x values of 0.5 mph, 1.0 mph, 5 mph, 10 mph, 25 mph, 50 mph, 60 mph, 75 mph, and 100 mph.
   b. Sketch a graph of the function $f(x)$.
   c. Identify any vertical asymptotes for the graph of the function $f(x)$, and interpret its meaning in the context of this problem.
   d. Identify any horizontal asymptotes for the graph of the function $f(x)$, and interpret its meaning in the context of this problem.
   e. State the domain and range of the function $f(x)$.
   f. Write an algebraic formula for $f(x)$.

4) The concentration of a drug in a person blood can often modeled nicely with a rational function. Suppose the following function gives the drug concentration $C(t)$ (in micrograms per mL) at the time t minutes after the drug was administered:

$$C(t) = \frac{5.1t}{0.01t^2 + 3.7}$$

   a. Sketch a graph of this function for t between 0 and 120 minutes.
   b. Find and interpret the roots, if any, of this function.
   c. Find and interpret the vertical asymptotes, if any, of this function.
   d. Find and interpret the horizontal asymptotes, if any, of this function.
   e. After approximately how many minutes is the drug concentration in the blood the highest?

5) Summarize how the end behavior of a function depends on the degrees of the numerator and denominator polynomials. In particular, determine the relationship between the degrees of the numerator polynomial and the denominator polynomial when:

a) The rational function has a horizontal asymptote at $y = 0$.

b) The rational function has a horizontal asymptote at some other y value (not zero).

c) The rational function tends to either positive or negative infinity it's end behavior.

Class Activity 27a: Paper to the Moon

Suppose you take a sheet of paper, cut it in half, and place one half on top of the other (so the stack is now 2 sheets high). Then you cut that stack in half, and place one half on top of the other (so the stack is now 4 sheets high). If you continue in this manner, how many cuts will it take until your stack reaches the moon?

Class Activity 27b: Graphs of Exponential and Log Functions

Your task is to sketch an accurate graph of each function given below.

1. $f(x) = 2^x$

2. $f(x) = \log_2 x$

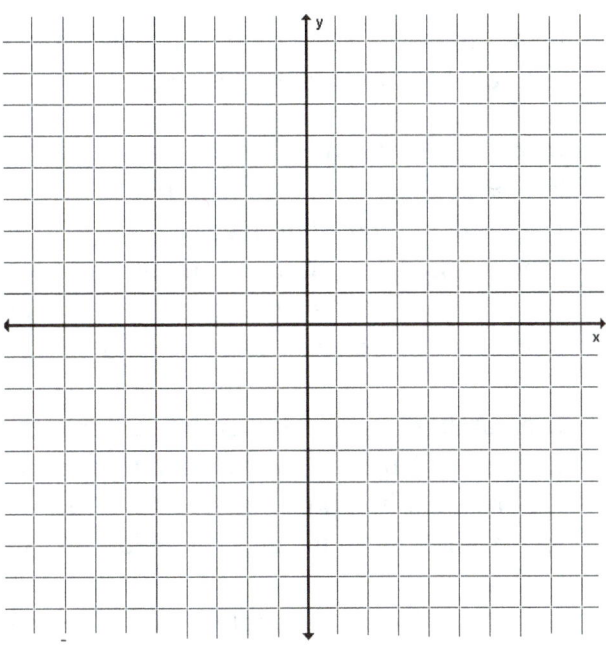

5. $f(x) = \left(\dfrac{1}{2}\right)^x$

6. $f(x) = \log_{\frac{1}{2}} x$

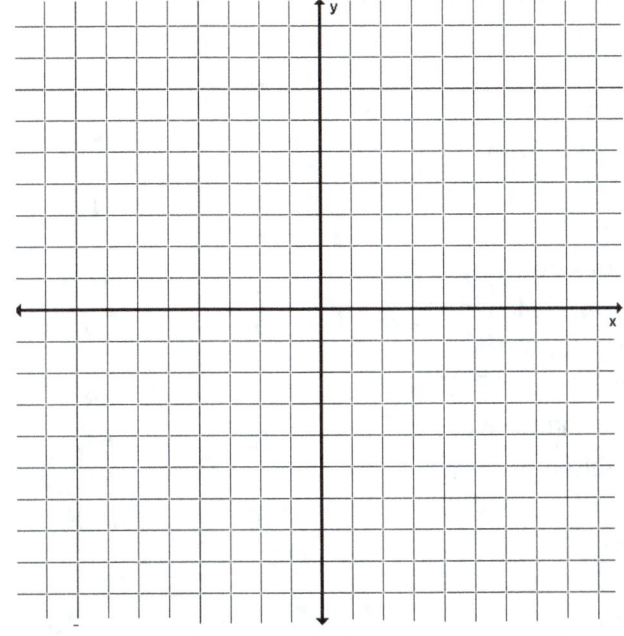

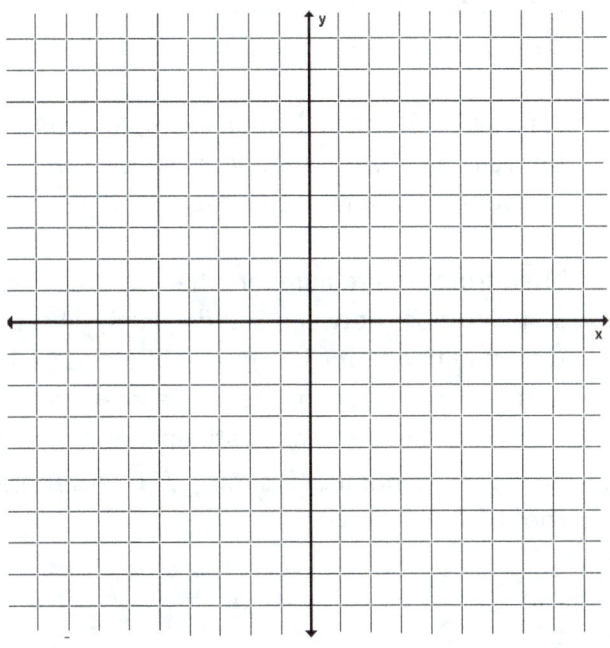

Read and Study 27: The Definition of the Logarithm Function

In the Paper to the Moon problem, we had an exponential function that gave us the height of a stack of paper as a function of the number of times we cut the paper. In order to find how many cuts it would take to get to the moon, we need to figure out what power of 2 is gives us that really large number. Let's reduce the problem somewhat. Suppose we want to solve the equation $2^x = 100$.

If we explore various powers of 2, we'll find that as $2^6 = 64$ and $2^7 = 128$, so that if $2^x = 100$, x should be some number between 6 and 7.

Using some trial and error and a calculator, we could compute $2^{6.5} \approx 90.5$, and then $2^{6.65} \approx 100.4$, to see that x is about 6.65. The exact number your need to raise 2 to the power of to get exactly 100 is an irrational number. Since it can't be written as a fraction, or indeed even as a root, we use a new notation to write the number: we call it $\log_2 100$.

The number $\log_2 100$ is just that, a number. It's not a problem to be solved, it's not a calculation to be done, it's just a number. That number is approximately 6.643856..., but that's just an approximation. So the best name for this number is $\log_2 100$.

x	2^x
0	1
1	2
2	4
3	8
4	16
5	32
6	64
7	128
8	256
9	512
10	1024

Sometimes there is a simpler way to represent logarithm. For example, suppose we wanted to solve the equation $2^x = 32$. Then $x = \log_2 32$. But there's a simpler name for the number $\log_2 32$, namely 5. Since 5 is the number that you raise 2 to to get 32, that means $\log_2 32$ is 5.

There is an important similarity between logarithms and roots. Both are undoing an exponent (or power). The only difference (and it's an important difference) is that with roots, we know the exponent, and want to know the number that's the base. With logarithms, we know the base, and want to know the number that's the exponent.

Let's go back and look at an example of finding a root and compare our reasoning with what we did above with finding a logarithm. Suppose we want to solve the equation $x^2 = 10$.

Then from square numbers that we know, we can figure out that x is some number between 3 and 4, likely closer to 3. Using a calculator, we could find that $(3.16)^2 \approx 9.986$ and $(3.17)^2 \approx 10.05$, so the number you square to get 10 is somewhere between 3.16 and 3.17. The exact number is an irrational number, and the best name we have for this number is simply $\sqrt{10}$. In general, we defined the nth root of x as follows:

x	x^2
0	0
1	1
2	4
3	9
4	16
5	25
6	36
7	49
8	64
9	81
10	100

$$y = \sqrt[n]{x} \text{ means } y^n = x.$$

So finding nth roots is the inverse process of finding nth powers. Similarly, finding logarithms for a base b is just the inverse process of finding exponentials with the base b. We define the logarithm function (with base b) to be the inverse of the exponential function (with base b) is called a logarithm function. That is:

$$y = \log_b x \text{ means } b^y = x.$$

Returning our attention to the first table showing the powers of 2, we can see lots of logarithms: every exponential equation can be rewritten as a logarithm equation. Since $2^3 = 8$, that means $\log_2 8 = 3$. Since $2^6 = 64$, that means $\log_2 64 = 6$.

This idea of an inverse is pervasive in algebra, and facility with inverse processes is another one of the "Algebraic Habits of Mind" that Mark Driscoll identifies in *Fostering Algebraic Thinking*. He calls this habit of mind **Doing/Undoing**, meaning that one process does something, and another related process undoes it. Often, we need to go back and forth (and back and forth), doing and undoing things when solving problems and thinking algebraically.

Some common examples inverse processes in algebra include:
- subtraction and addition undo each other
- division and multiplication undo each other
- roots and powers undo each other
- logarithms and exponents undo each other

If we view the equations in the definition of the logarithm as functions, then we say that the logarithm function is the **inverse function** for an exponential function. We will make a formal definition of inverse functions in a later section, but in short, Inverse functions result from switching the roles of x and y, in other words, of switching what's considered the input and the output of the function.

Homework Set 27

1) Evaluate the following, without using any calculator logarithm buttons.

a. $\log_3 81$
b. $\log_3(9)$
c. $\log_9(3)$
d. $\log_9\left(\frac{1}{3}\right)$
e. $\log_9(27)$
f. $\log_3(\sqrt{3})$

g. $\log_3 \frac{1}{27}$
h. $\log_{64} 8$
i. $\log_{64} 4$
j. $\log_8 2$
k. $\log_8 4$

2) Sketch an accurate graph of the following functions on a 10 x 10 grid. Do this by plotting points that you can calculate by hand. Do not use a graphing calculator.

   a. $f(x) = 3^x$
   b. $f(x) = \log_3 x$
   c. $f(x) = \left(\frac{3}{2}\right)^x$
   d. $f(x) = \log_{\frac{3}{2}} x$
   e. $f(x) = \left(\frac{2}{3}\right)^x$
   f. $f(x) = \log_{\frac{2}{3}} x$

3) Based on your results from the second class activity and from the previous homework problem, summarize the key characteristics of the given exponential and logarithm function forms, by specifying:

   - roots (if any)
   - y-intercept (if any)
   - end behavior as $x \to -\infty$
   - end behavior as $x \to +\infty$
   - vertical asymptotes (if any)
   - domain
   - range

   a) The exponential function $f(x) = b^x$, where $b > 1$
   b) The exponential function $f(x) = b^x$, where $0 < b < 1$
   c) The logarithm function $f(x) = \log_b x$, where $b > 1$
   d) The exponential function $f(x) = \log_b x$, where $0 < b < 1$.

4) Find exact solutions to the following equations.
   a) $2^{x-1} = 16$
   b) $2^{x-1} = 5$
   c) $3^{4x+1} = 1$
   d) $3^{4x+1} = 10$
   e) $3^{(1-3x)} = 3^{(x+9)}$
   f) $9^{6-x} = 3^x$
   g) $4 \cdot 2^{3x} = \left(\frac{1}{2}\right)^x$

5) Find exact solutions to the following equations.
   a) $\log_5 x = {}^-1$
   b) $\log_{10}(4x) = 4$
   c) $\log_4(x - 5) = 0$
   d) $\log_4(x) - 5 = 0$
   e) $2 \log_9(3x - 1) = 1$
   f) $\log_2\left(\frac{x-1}{3x+4}\right)$

6) Determine whether the following are valid property of logarithms. Support your answer with numerical examples.
   a) $\log_b(u + v) = \log_b(u) + \log_b(v)$
   b) $\log_b(uv) = \log_b(u) + \log_b(v)$
   c) $\log_b(u^n) = n \cdot \log_b(u)$

7) Suppose a strain of bacteria in a nutrient-rich environment grows so that the amount triples every hour. Suppose you start at $t = 0$ with sample of 5 micrograms of this bacteria.

   a. Let t number of hours that the sample has been growing, and let $f(t)$ be the number of micrograms of bacteria in the sample as a function of t. Write a formula for $f(t)$.

   b. How many micrograms of bacteria will there be at $t = 4$ hours?

   c. How many micrograms of bacteria will there be at $t = \frac{1}{2}$ hour? Give an exact answer, as well as a decimal approximation to the nearest tenth.

   d. After how many hours will there be exactly 10,000 micrograms of bacteria? Give an exact answer, as well as a decimal approximation to the nearest tenth.

Class Activity 28a: Properties of Logarithms

1. Simplify the following expressions:

    a. $\log_2(8^5)$

    b. $\log_2(8^n)$

    c. $\log_2[(2^m)^n]$

2. Derive a property of logarithms that will rewrite the expression $\log_b(u^n)$.
   Start with $\log_b(u^n)$, then substitute $u = b^m$, and simplify like we did above.

3. Simplify the following expressions:

    a. $\log_2(8 \cdot 32)$

    b. $\log_3(9\sqrt{3})$

4. Derive a property of logarithms that will rewrite the expression $\log_b(u \cdot v)$.
   Start with $\log_b(u \cdot v)$, then substitute $u = b^m$, and $v = b^n$ simplify like we did above.

Class Activity 28b: Growing All the Time

Suppose you have one unit of something. Let's say you have 1 liter of water. Wait, that doesn't seem very valuable. Ok, how about 1 pound of gold.

If you had 1 pound of gold, and if somehow you were able to increase the amount you had by 100%, then your amount would double to 2 pounds, right?

What if instead, starting again with 1 pound of gold, you increased the amount you had by 50%, then increased this amount by 50%. In other words, your increase your original amount by 50% twice. Would you still end of with 2 pounds of gold? Explain.

What if you start with 1, then increase by 25% four times?

What if you start with 1, then increase by 20% five times?

What if you start with 1, then increase by 10% ten times?

What if you start with 1, then increase by 5% twenty times?

What if you start with 1, then increase by 1% one hundred times?

Find a formula for increasing your amount by $\left(\frac{100}{n}\right)$% a total of n times.

What happens to this process as n approaches infinity?

Read and Study 28: Properties of Logarithms and the Natural Base *e*

When you are through changing, you are through.

Bruce Barton

In the "Growing All the Time" activity, we saw that a special number occurs when growth is taking place continuously. Specifically, the number e, also called Euler's constant, or the natural exponent base, is defined as the limit as n tends to infinity of the quantity $(1 + 1/n)^n$. In the activity, we thought of this quantity as the result of starting with the number 1 and repeatedly by increasing by $\frac{1}{n}$ a total of n times. First we note that to increase by $\frac{1}{n}$ is equivalent to multiplying by $(1 + \frac{1}{n})$. For example, if we want to increase a quantity by $\frac{1}{2}$ we'd multiply by $(1 + \frac{1}{2})$. To increase by $\frac{1}{2}$ twice, we'd multiply by $(1 + \frac{1}{2})$ then again by $(1 + \frac{1}{2})$ so really we've multiplied by $(1 + \frac{1}{2})^2$. Similarly if we want to increase a quantity by $\frac{1}{10}$, we'd multiply by $(1 + \frac{1}{10})$. To increase by $\frac{1}{10}$ ten times, we'd multiply by $(1 + \frac{1}{10})^{10}$.

So if we increase by $\frac{1}{n}$ a total of n times, that is the same as multiplying by $(1 + \frac{1}{n})^n$. In the following table we will show the value of $(1 + \frac{1}{n})^n$ as n gets larger.

n	$\left(1 + \dfrac{1}{n}\right)^n$	Decimal or decimal approximation
1	$\left(1 + \dfrac{1}{1}\right)^1$	2
2	$\left(1 + \dfrac{1}{2}\right)^2$	2.25
3	$\left(1 + \dfrac{1}{3}\right)^3$	2.37037037…
4	$\left(1 + \dfrac{1}{4}\right)^4$	2.44140625
5	$\left(1 + \dfrac{1}{5}\right)^5$	2.48832
6	$\left(1 + \dfrac{1}{6}\right)^6$	2.521626372…
7	$\left(1 + \dfrac{1}{7}\right)^7$	2.546499697…
8	$\left(1 + \dfrac{1}{8}\right)^8$	2.565784514…
9	$\left(1 + \dfrac{1}{9}\right)^9$	2.581174792…
10	$\left(1 + \dfrac{1}{10}\right)^{10}$	2.59374246…

The values are growing, but slowly. Will they continue to get larger and larger? Will they ever get above 3? Let's look at more calculations:

n	$\left(1+\dfrac{1}{n}\right)^n$	Decimal or decimal approximation
50	$\left(1+\dfrac{1}{50}\right)^{50}$	2.691588029…
100	$\left(1+\dfrac{1}{100}\right)^{100}$	2.704813829…
250	$\left(1+\dfrac{1}{250}\right)^{250}$	2.712865123…
500	$\left(1+\dfrac{1}{500}\right)^{500}$	2.715568521…
1000	$\left(1+\dfrac{1}{1000}\right)^{1000}$	2.716923932…
2500	$\left(1+\dfrac{1}{2500}\right)^{2500}$	2.717738371…
5000	$\left(1+\dfrac{1}{5000}\right)^{5000}$	2.718010050…
10,000	$\left(1+\dfrac{1}{10,000}\right)^{10000}$	2.718145927…
100,000	$\left(1+\dfrac{1}{100,000}\right)^{100,000}$	2.718268237…
1,000,000	$\left(1+\dfrac{1}{1,000,000}\right)^{1,000,000}$	2.718280469…

The limiting value as *n* tends to infinity is an irrational number, meaning it can't be written as a fraction of integers. But it also can't be written as some root (square root, cube root, etc.) of any rational number either. That means we have no way of writing this number exactly, other than giving it its own symbol, and the symbol that's used is *e*. (Another famous number that's like this is π.)

The number *e* has lots of very cool properties, and it makes a very convenient base for an exponential function, $f(x) = e^x$. Even though bases such as 2 or 10 are more intuitive to think about, since we are used to doubling and multiplying by 10, using base e is actually more convenient in many applications. If you study calculus later, you will come to appreciate the usefulness of working with the exponential and logarithm functions with base *e*.

We should also discuss some commonly used notation:

$$\ln(\ ) \text{ means } \log_e(\ ).$$
$$\log(\ ) \text{ means } \log_{10}(\ ).$$

The "ln" is short for "logarithm natural", and $f(x) = \ln x$ is called the natural logarithm function. The description "natural" comes from the fact that the number e arises out of this naturally occurring growth pattern shown in the tables above. And since the logarithm was first invented to deal with powers of 10, the notation log( ) where a base isn't specified is assumed to mean the base is 10.

In the first class activity, you derived two properties of logarithms. We list these, and also a third, which we will ask you to prove in the homework, in the box below. Notice how each of these properties of logarithms just the same as one of the properties of exponents, just rewritten in logarithm form. For example, the first property says the exponent of something raised to an exponent is the product of exponents. The second property says that the exponent of a product is the sum of the exponents. And the third property says the exponent of a quotient is the difference of the exponents.

Key Properties of Exponents and their Corresponding Properties of Logarithms

1) $(b^m)^n = b^{m \cdot n}$ corresponds to $\log_b(u^n) = n \cdot \log_b(u)$

2) $b^m \cdot b^n = b^{m+n}$ corresponds to $\log_b(u \cdot v) = \log_b(u) + \log_b(v)$

3) $\dfrac{b^m}{b^n} = b^{m-n}$ corresponds to $\log_b\left(\dfrac{u}{v}\right) = \log_b(u) - \log_b(v)$

Again, we are listing these properties to make it easier for us to talk about them, not for you to simply memorize. What's important is that you can understand why these properties make sense, and recognize when you might want to make used of them. In particular, notice that $\log_b(u) + \log_b(v)$ is not equivalent to $\log_b(u + v)$, but instead $\log_b(u \cdot v)$. Remember that the distributive law is **only** for multiplication over addition: logarithms don't distribute, just like exponents don't distribute, roots don't distribute, inverses don't distribute, etc.

A common method used to solve an equation for a variable that is inside an exponent is to "take the logarithm" of both sides of the equation and use property 1 listed above to rewrite the expression with the variable as a factor, rather than an exponent. In fact, since this property works regardless of what the base b is for the logarithm, its common to just use a familiar base such as base ten or base e.

For example, suppose you have the equation $3^x = 50$, and you wanted to solve for x. Well, we know that by definition that $x = \log_3 50$. So there's really nothing to do to solve the equation, just rewrite it using the definition of the logarithm:

	3^x	=	50	
	x	=	$\log_3(50)$	By definition of the logarithm

Suppose now you want to get a decimal approximation for this number $\log_3 50$, but the calculator you have only has buttons for $\log_{10}(\ )$ or $\ln(\ )$. Then one method would be to input both 3^x and 50 into one of these logarithm functions and solve the resulting equation for x.

	3^x	=	50	
$\log_{10}(3^x)$	=	$\log_{10}(50)$	Input both sides into the $\log_{10}$ function ("take the logarithm" of both sides)
$x \log_{10}(3)$	=	$\log_{10}(50)$	Property 1 of Logarithms
x	=	$\dfrac{\log_{10}(50)}{\log_{10}(3)}$	Divide both sides by $\log_{10}(3)$

Now this expression is rather messy, and its only advantage is if you can calculate logs base ten, but not logs base three. Similarly, one could solve using the natural logarithm:

	3^x	=	50	
$\ln(3^x)$	=	$\ln(50)$	Input both sides into the natural logarithm
$x \ln(3)$	=	$\ln(50)$	Property 1 of Logarithms
x	=	$\dfrac{\ln(50)}{\ln(3)}$	Divide both sides by $\ln(3)$

Homework Set 28

1) Evaluate the following, without using a calculator.
    a) $\log_{17}(17)$
    b) $\ln(e^3)$
    c) $\log(10^4)$
    d) $\ln(e)$
    e) $\ln\left(\dfrac{1}{e}\right)$
    f) $\log(10000)$
    g) $\log(0.01)$

2) Find exact solutions to the following equations.
   a) $e^{3x+1} = 1$
   b) $3e^x - 7 = 5$
   c) $20e^{-2x} = 5$
   d) $\ln(2x - 3) = 0$
   e) $\log(4x - 2) = 1$
   f) $\ln(\sqrt{x}) = {-}1$

3) Below are some of the exponential equations you solved in the previous section. Now try solving them by the "taking the logarithm of both sides" method and using the properties of logarithms we derived in this section.

   a) $3^{4x+1} = 1$
   b) $3^{4x+1} = 10$
   c) $3^{(1-3x)} = 3^{(x+9)}$
   d) $9^{6-x} = 3^x$
   e) $4 \cdot 2^{3x} = \left(\frac{1}{2}\right)^x$

4) Solve the following equations without using property 1. Show that they give the same solution.

   a) $\log_3(x^2) = 3$
   b) $2\log_3 x = 3$

5) Find exact solutions to the following equations. Suggestion: first apply the appropriate exponential function to both sides and Property 2 or 3 of exponents. Alternately, use the corresponding Property 2 or 3 of logarithms to rewrite the equation so that it has only one logarithm. Better yet, try it both ways and see that you get the same result!

   a) $\log(x) + \log(4x) = 2$
   b) $\log(x + 1) - \log(3x - 4) = {-}1$
   c) $\log(2x) = \log(x - 1) + 1$

6) The Arrhenius equation, $k = Ae^{-\frac{E}{RT}}$, gives the rate constant k of a chemical reaction as a function of the absolute temperature T, the frequency factor A, the activation energy E, and the universal gas constant R. Solve the Arrhenius equation for the activation energy E.

7) Suppose you invest $1000 in retirement account at 5% annual interest. That means after 1 year you will have $1050 in this account, and after 2 years you will have $1102.50.
   a) Explain why adding 5% to an amount can be calculated by multiplying that amount by 1.05.
   b) What would be the value in this account after 10 years? 25 years? 50 years?
   c) How long will it take the original investment of $1000 to double to $2000? How long will it take the original investment of $1000 to quadruple to $4000?
   d) Write an algebraic formula for the amount in the account as a function of years.
   e) Sketch a graph of this function.

8) The half-life of a radioactive substance is the amount of time required for half of an amount of the substance to decay. The half-life of argon-37 is about 35 days. Suppose you start with an 18-gram sample of argon-37. Then after 35 days there will be only 9 grams left, and after another 35 days, there will be only 4.5 grams left.
   a) Make a table that shows how much of this 18-gram sample is remaining after 0, 35, 70, 105, 140, 175 and 210 days.
   b) How much argon-37 will be remaining after exactly 17.5 days? Express this amount as an exact number and as a decimal approximation accurate to 1 decimal place.
   c) After how many days will there be exactly 2 grams remaining? Express this amount as an exact number and as a decimal approximation accurate to 1 decimal place.
   d) Let $f(t)$ be the amount of argon-37 remaining after t days. Write a formula for $f(t)$.
   e) Use the values you computed in parts a., b., and c. to sketch a graph of the function $f(t)$ on grid paper.

9) Newton's Law of Heating and Cooling states that an object will cool or warm up at a rate proportional to the difference between its temperature and the temperature of its surroundings. A function that models this behavior is the following exponential function.

$$T(t) = T_a + (T_0 - T_a)e^{-kt}$$

where T_a is the ambient (surrounding) temperature, T_0 is the initial temperature of the object, and k is a constant rate determined by the object (different materials cool down and heat up at different rates).

   a) Verify that the formula gives $T(0) = T_0$.
   b) What happens to the value of $T(t)$ as t tends to infinity?

10) Suppose you have a pot of soup you took just off the stove at a temperature of 212 degrees Fahrenheit, and you let it sit in a 72 degree room to cool. After 5 minutes the temperature is 180 degrees.
   a) Use Newton's Law of Heating and Cooling in the previous problem to approximate the cooling rate k for this soup, then write a function for the temperature of this soup as a function of time.
   b) Use your function to determine the temperature of the soup after 10 minutes.
   c) Use your function to determine how long you have to wait until the temperature of the soup will be 125 degrees.
   d) Use your function to determine how long you have to wait until the temperature of the soup will be 73 degrees.
   Use the data you now have to sketch a graph of the temperature of this soup as a function of time.

Class Activity 29a: Back and Forth

Recall the Crossing the River problem.

1. Let x be the number of adults, and let $f(x)$ be the number of trips required to get everyone across the river.

   a. Find $f(25)$ and interpret its meaning.
   b. Find a formula for $f(x)$.
   c. Make a graph of the function $f(x)$.

2. Now let x be the number of trips required to get everyone across, and let $g(x)$ be the number of adults in the group.

   a. Find $g(25)$ and interpret its meaning.
   b. Find a formula for $g(x)$.
   c. Make a graph of the function $g(x)$.

3. How are the graphs of $f(x)$ and $g(x)$ related?

4. Figure out the value of the following:

   a. $f(g(45))$
   b. $g(f(30))$
   c. $f(g(x))$
   d. $g(f(x))$

Class Activity 29b: Inverse Function Machines

Below is a function defined by a sequence of operations as labeled.

a. Write a formula for $f(x)$ and determine its domain.
b. Determine which **input(s)** would result in the given output.
c. Find an expression for the input(s) into the function would result in the **output of y**.
d. Make a new function machine diagram that reverses this process. Make your machine go from right to left.
e. If you now consider y to be the **input** into the inverse process, is this inverse process a function of y? Explain.
f. If the inverse is a function, write its formula, and state its domain.
g. Determine the range of the function $f(x)$.

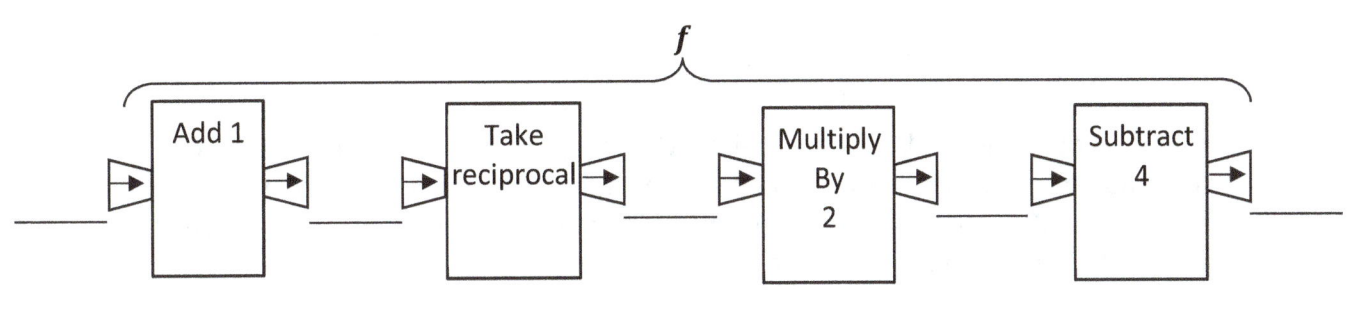

$f(x) =$ \hspace{2em} Domain of f:

$f() = 4$ \hspace{4em} $f() = y$

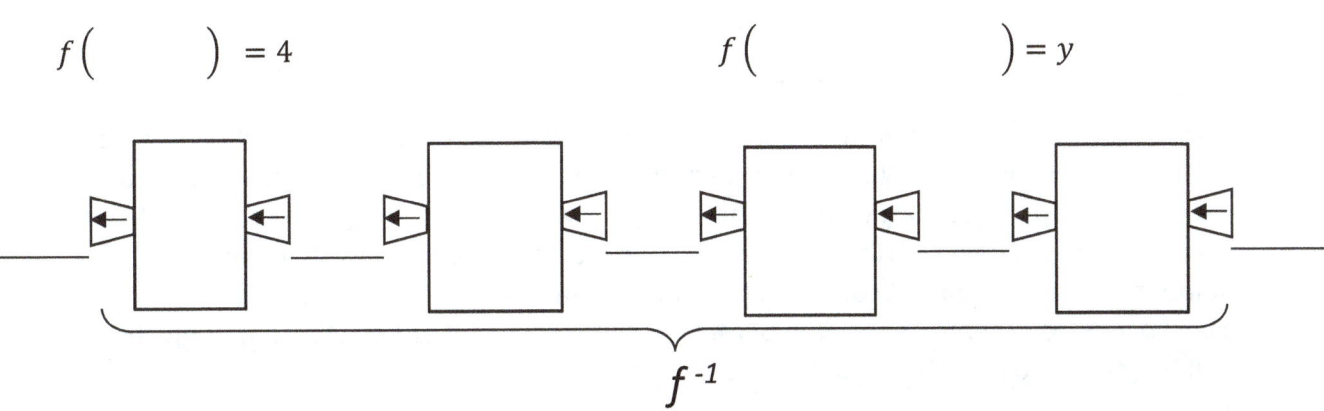

$f^{-1}(y) =$ \hspace{2em} Domain of f^{-1}:

\hspace{8em} Range of f:

Read and Study 29: Inverse Function Definition and Notation

It takes me maybe an hour to understand a page of new mathematics. At the end, if I'm lucky I obviously say to myself "Why didn't I understand that right away? What took the hour?"
— Marvin Minsky

As we discussed in the first chapter, a function can be thought of as a process that takes an input and ultimately returns a unique output. We can also think about reversing that process, starting with that output and finding the original input. If that reverse process is also a function, then we call it the inverse function.

Let's explore this idea with a specific example. Consider the function $f(x) = 2 - 3x$. This function takes an input of x and returns an output of $2 - 3x$.

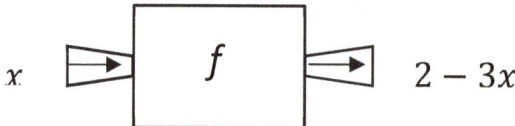

So what goes on inside the box? If we look at what happens to the input number x, first the function multiplies this number by 3, then subtracts the result from 2. For example, if the input were 5, then multiplying by 3 gives 15. Then subtracting this from 2 gives $^-13$. So we could write a function machine diagram for f with two steps, as shown:

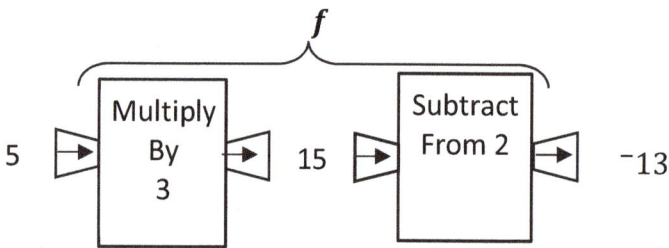

Now if we want instead to write a function for the *reverse* process, the one that takes an input of $^-13$, undoes these steps to get an output of 5, we just need to think about what undoes each of the steps. If we start with $^-13$, how do you get 15? Hmm. It's not obvious what the computation is that undoes "subtract from 2". So let's break it down further. Subtracting a number, as we will explore more in later sections, is just adding the number that is opposite in sign. So instead of subtracting a number from 2, let's think of it as two steps: multiply the number by negative 1, then add this to 2.

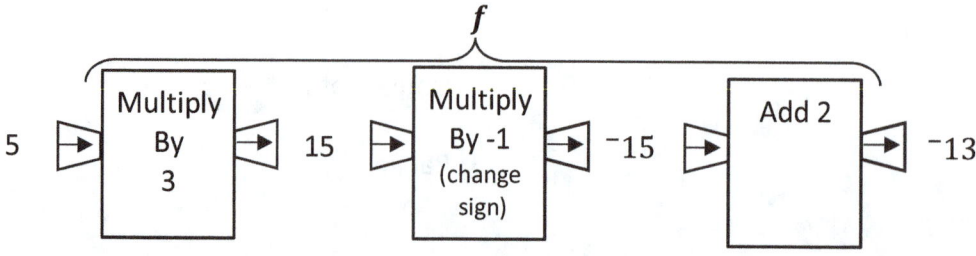

Let's try to find the steps that undoes each of these. If we start with ⁻13, we can undo "adding 2" by "subtracting 2" to get ⁻15. Then we can undo "multiply by ⁻1" by "multiplying by ⁻1". (Isn't that neat? Multiplying by ⁻1 is the same as changing the sign of a number. And how do you undo changing the sign of a number? Change the sign again!). So we get 15. Then we can undo "Multiply by 3" by "Dividing by 3" to get 5. So here's the new function machine, relabeled so we now go from right to left).

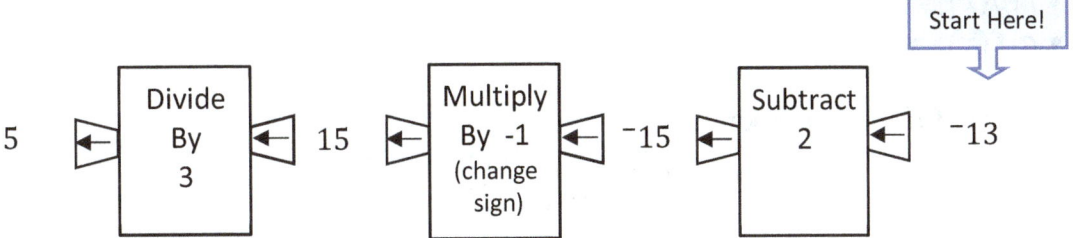

Now let's see what happens if we send x back through the machine. (Read this diagram from right to left.)

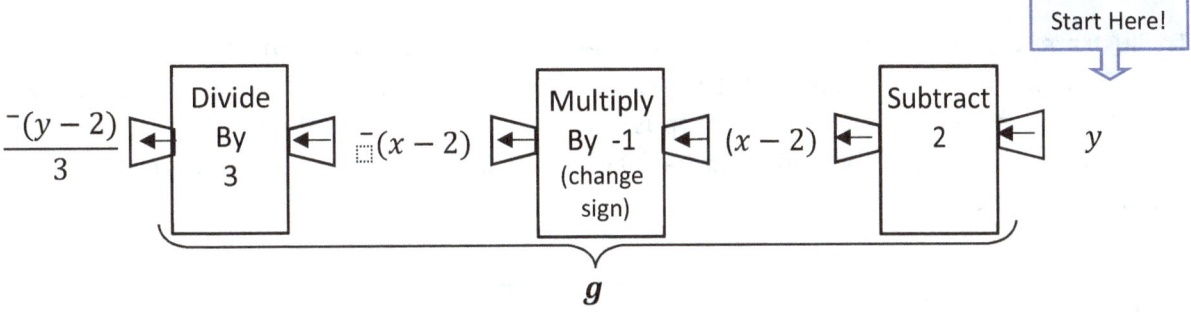

If we read this machine diagram from right to left we can consider this to be a new function, we could call it $g(y)$. Since this function takes an input of y and returns an output of $\frac{^-(y-2)}{3}$, a formula for this function would be $g(y) = \frac{^-(y-2)}{3}$.

Now here's the thing. Previously, we said that it is standard to use x for the variable that is the input into a function, and to use y for the output. But when we write $g(y) = \frac{^-(y-2)}{3}$ we are violating this convention. But really we could use any symbol to represent the variable. In fact, we could write the function g as

$$g(y) = \frac{^-(y-2)}{3}, \text{ or}$$
$$g(x) = \frac{^-(x-2)}{3}, \text{ or}$$
$$g(t) = \frac{^-(t-2)}{3}, \text{ or even}$$
$$g(\text{Fred}) = \frac{^-(\text{Fred}-2)}{3},$$

assuming that "Fred" was a number. So we will just call the function $g(x) = \frac{^-(x-2)}{3}$.

This function $g(x)$ that undoes the function $f(x)$ is called the **inverse** of the function f, and we can write $g(x) = f^{-1}(x)$. Here is the definition:

Two functions $f(x)$ and $g(x)$ are said to be **inverse functions** of each other provided:

- $f[g(x)] = x$, and
- $g[f(x)] = x$.

The **inverse function notation:**
$$g(x) = f^{-1}(x), \quad \text{and}$$
$$f(x) = g^{-1}(x)$$

mean that $f(x)$ and $g(x)$ are inverse functions of each other.

To illustrate, we will show that the functions f and g in our example fit this definition for inverse functions. First let's show that $f[g(x)] = x$. What this means is that if the number x is put into the function g, and then the output number $g(x)$ is put into the function f, then the result will be that we get the number x back again. We'll say that again, since it is crucial that you understand this:

The equation $f[g(x)] = x$ means that if the number x is put into the function g, and then the output number $g(x)$ is put into the function f, then we will get the original number x back again.

Well, we already know that if the number x is put into the function g, that the output will be the number $\frac{-(x-2)}{3}$. That's what it means when we defined $g(x) = \frac{-(x-2)}{3}$. So now let's take the number $\frac{-(x-2)}{3}$ and put it into the function f. We can use the function machine diagram for f to illustrate:

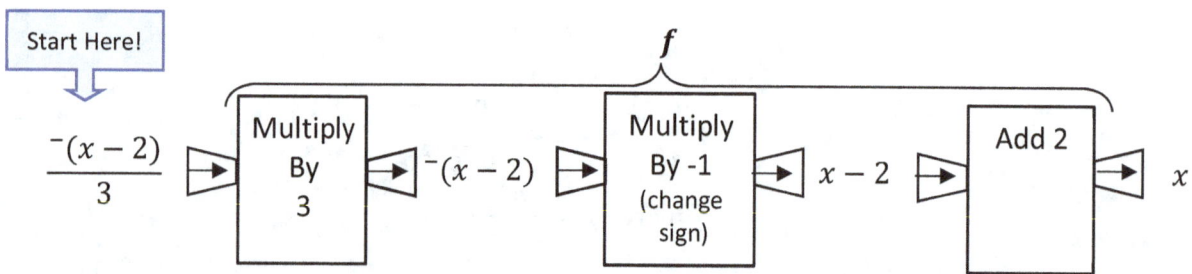

218

Make sure each of step of the diagram makes sense and that the outputs are correct. Notice that the inputs and outputs are identical between this machine diagram for f (which goes left to right), and the last diagram for g on the previous page (which went right to left), which is of course by design!

Or, without using the machine diagram, we could write:

$$\begin{aligned} f[g(x)] &= f\left[\frac{^-(x-2)}{3}\right] \\ &= 2 - 3 \cdot \left[\frac{^-(x-2)}{3}\right] \\ &= 2 - {^-(x-2)} \\ &= 2 + x - 2 \\ &= x \end{aligned}$$

This way of writing the calculation is a little more abstract, but it represents the same process as in the machine diagram. Compare the methods carefully and see how each step in the machine diagram corresponds to a step in simplifying the expression $f[g(x)]$.

Similarly, let's now show in our example that $g[f(x)] = x$. This means that if the number x is put into the function f, and then the output number $f(x)$ is put into the function g, then we will get the number x back again. We already have an expression for the output when x is put into the function f, namely $f(x) = 2 - 3x$. So now let's put that output into the function g. Using the machine diagram for g, (we will keep it going from right to left):

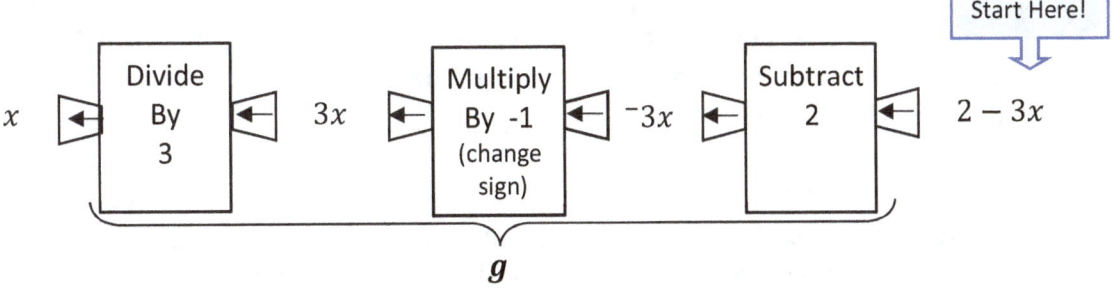

Or, without using the machine diagram, we could write:

$$\begin{aligned} g[f(x)] &= g[2 - 3x] \\ &= \frac{^-([2-3x] - 2)}{3} \\ &= \frac{^-(^-3x)}{3} \\ &= \frac{3x}{3} \\ &= x \end{aligned}$$

Again, you should compare the two methods step by step.

Since we have shown that $f[g(x)] = x$, and that $g[f(x)] = x$, this means that by definition, the two functions in our example are inverses of each other. There are several ways we could state this fact. The following are equivalent:

- If $f(x) = 2 - 3x$, then $f^{-1}(x) = \frac{^-x+2}{3}$.
- If $g(x) = \frac{^-x+2}{3}$, then $g^{-1}(x) = 2 - 3x$.
- The functions $f(x) = 2 - 3x$ and $g(x) = \frac{^-x+2}{3}$ are inverses of each other.

In the homework, we will give you more pairs of functions to verify in this way whether they are inverses of each other.

More on Inverses. While the inverse function notation looks like an exponent of negative one, it is not the same as an exponent. Exponents are superscripts above *numbers*, and indicate a repeated multiplication. In particular, an *exponent* of $^-1$ indicates a reciprocal. That is,

If x is a number, then x^{-1} means $\frac{1}{x}$.

However, in the inverse function notation, the $^-1$ is a superscript over a *function*, not a number. So it is not an exponent, but just an indication of the inverse.

If f is a function, then f^{-1} means the inverse function of f.

In particular, $f^{-1}(x)$ does NOT mean $\frac{1}{f(x)}$. For example, note that the inverse of the function $f(x) = 2 - 3x$ is not the function $\frac{1}{2-3x}$. In the exercises, we will have you prove this directly that these are not inverse functions. (And we just finished proving that $g(x) = \frac{^-x+2}{3}$ is in fact the correct inverse function for f.)

So if inverse function notation does not mean a reciprocal, why is the same notation used for both inverse functions and reciprocals? The reason is that in both cases the $^-1$ superscript denotes an inverse. With exponents, which are defined in terms of multiplication, the $^-1$ exponent means the **multiplicative inverse,** and the multiplicative inverse of a number turns out to be its reciprocal. So this is just a case of people using the same notation to denote inverses of functions as they do for inverses of multiplication.

Recall that **multiplicative inverses** are numbers that multiply together to get to the **multiplicative identity**, which is the number 1. For example, a and $\frac{1}{a}$ are multiplicative inverses since they multiply to get 1, which is the identity for multiplication.

$$a \cdot \frac{1}{a} = 1$$

Similarly, **additive inverses** are numbers that add together to get to the **additive identity**, which is the number 0. For example, a and ^-a are additive inverses since they add together to get 0, which is the identity for addition

$$a + {}^-a = 0$$

Now we have **function inverses** are functions that compose together to the **identity function**, which is $y = x$. (In the homework, we will ask you to think about why it makes sense to say that $y = x$ is the identity function):

$$f[f^{-1}(x)] = x$$

The mathematician Henri Poincare once said that mathematics is the art of giving the same name to different things. The names "identity" and "inverse" and are prime examples of this. The number 0, the number 1, and the function $y = x$ are all different things, but they are all identity elements. Similarly the number $\frac{1}{a}$, the number ^-a , and the function $y = f^{-1}(x)$ are all different things, but they are all inverses for something under some operation.

Homework Set 29

1) Consider the following invertible function: $f(x) = x^3 + 5x + 3$. Find the value of the following:

   a) $f(3)$
   b) $f^{-1}(45)$
   c) $f(0)$
   d) $f^{-1}(3)$

   e) $f^{-1}(9)$
   f) $f(^-3)$
   g) $f^{-1}(^-3)$

   (Note: you do not need to find a formula for $f^{-1}(x)$ to be able to determine these values. A little investigation and experimentation is sufficient).

2) Consider the function $f(x) = 2x^3 - 5$.
   a) Make a machine diagram for $f(x)$.
   b) Find $f(0)$ and $f^{-1}(0)$.
   c) Find $f(^-3)$ and $f^{-1}(^-3)$.
   d) Make a machine diagram for $f^{-1}(x)$, and write a formula for this function.
   e) Sketch a graph of $f(x)$ and state its domain and range.
   f) Sketch a graph of $f^{-1}(x)$ and state its domain and range.

3) Consider the function $f(x) = {}^-4 \cdot \left(\frac{1}{3x} + 2\right)$.
   a) Make a machine diagram for $f(x)$.
   b) Find $f(1)$ and $f^{-1}(1)$.
   c) Make a machine diagram for $f^{-1}(x)$, and write a formula for this function.
   d) Sketch a graph of $f(x)$ and state its domain and range.
   e) Sketch a graph of $f^{-1}(x)$ and state its domain and range.

4) If f and g are inverse functions, then $g[f(x)]$ should be x, and $f[g(x)]$ should be x. Use this idea to determine whether the following pairs of functions are inverses.
   a) $f(x) = 4x - 10$; $g(x) = \frac{x+10}{4}$
   b) $f(x) = 2 - 3x$; $g(x) = \frac{1}{2-3x}$.

5) Recall the railway bridge truss problem from Class Activity 19.
   a) How long is a truss that uses 75 beams?
   b) Write a formula for the length of a truss that uses n beams.
   c) Write a formula for the number of beams required to make a truss that has length n.
   d) Show that the formulas in part c and d represent a pair of inverse functions.

6) In Class Activity 29, we investigated the function $f(x) = \frac{2}{x+1} - 4$ and found that its inverse function was $f^{-1}(x) = \frac{2}{x+4} - 1$.
   a) Verify that $f^{-1}[f(x)] = x$.
   b) Sketch the graphs of the two functions $f(x) = \frac{2}{x+1} - 4$ and $f^{-1}(x) = \frac{2}{x+4} - 1$, on the same axes. Use tracing paper to demonstrate for yourself that if you flip the graphs so that the x and y axes get switched, that the graphs of f and f^{-1} switch.

7) Suppose f and g are inverse functions of each other. How are the domain and range of f and g related? What can you say about the geometric relationship between the graphs of a function and its inverse function? Why does this make sense?

8) In the Read and Study, we told you that the identity function is $f(x) = x$.
   a) Find the value of the following: $f(3)$, $f(^-12)$, $f\left(\frac{4}{7}\right)$, $f(\sqrt{10})$
   b) Sketch a graph of this function.
   c) Let $g(x)$ be any function. Find a formula for $f(g(x))$.
   d) In what way does the function $f(x) = x$ behave like the number 0 and the number 1? In other words, why does it make sense to call this the identity function?

Class Activity 30: More Inverse Function Machines

Below is a function defined by a sequence of operations as labeled.

a. Write a formula for $f(x)$ and determine its domain.
b. Determine which **input(s)** would result in the given output.
c. Find an expression for the input(s) into the function would result in an **output of y**.
d. Make a new function machine diagram that reverses this process. Make your machine go from right to left.
e. If you now consider y to be the **input** into the inverse process, is this inverse process a function of y? Explain.
f. If the inverse is a function, write its formula, and state its domain.
g. Determine the range of the function $f(x)$.

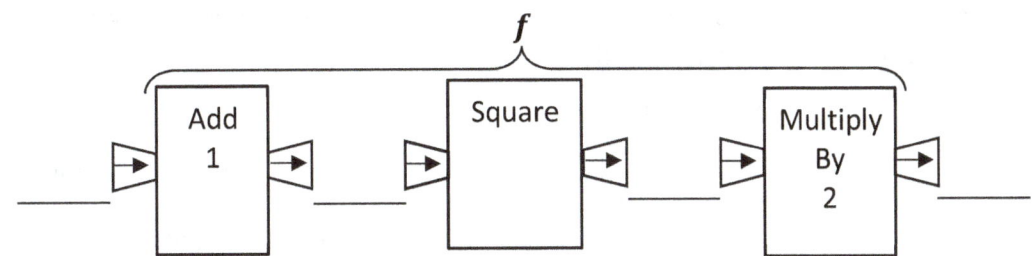

$f(x) =$ 	Domain of f:

$f\big(\big) = 10$ 	$f\big(\big) = y$

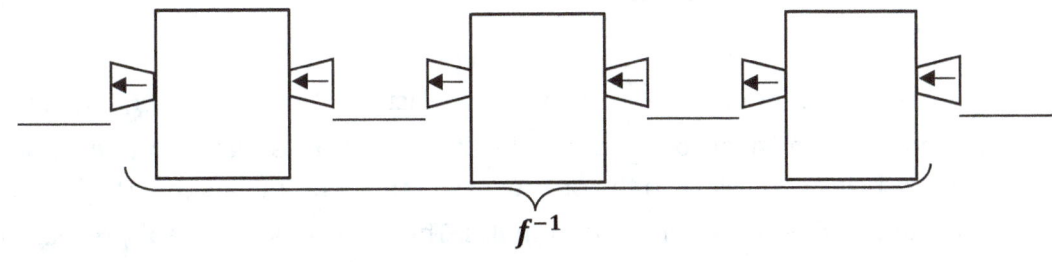

$f^{-1}(y) =$ 	Domain of f^{-1}:

Range of f:

223

Read and Study 30: Finding Inverse Function Formulas

Math is like love – a simple idea but it can get complicated.

R. Drabek

In the previous section, we thoroughly explored the example with the function $f(x) = 2 - 3x$. By reasoning backwards through the function machine, we determined that $f^{-1}(y) = \frac{^-(y-2)}{3}$ was the formula for its inverse function. That is, since the function f multiplied a number by negative 3, then added two, the inverse function will undo these steps, by first subtracting 2 then dividing by negative 3.

Another way to derive this inverse function formula would be to use our equation solving skills, as we will now demonstrate. Instead of using function notation, we will just use the variable x for the input into f (and the output of f^{-1}) and the variable y for the output of f (and the input of f^{-1}). We will start by writing the formula for the function f, then solve the equation for x.

$$y = f(x)$$

$$y = 2 - 3x \quad \text{Substitute the formula for } f$$

$$y + 3x = 2 - 3x + 3x \quad \text{Property of Equality: Add 3x to both sides}$$

$$y + 3x = 2 \quad \text{Since } {}^-3x + 3x = 0$$

$$y + 3x - y = 2 - y \quad \text{Property of Equality: Add } {}^-y \text{ to both sides}$$

$$3x = 2 - y \quad \text{Since } y - y = 0$$

$$\frac{3x}{3} = \frac{2-y}{3} \quad \text{Property of Equality: Multiply } \frac{1}{3} \text{ to both sides,}$$

$$x = \frac{2-y}{3} \quad \text{Since } \frac{3}{3} = 1$$

Notice that last equation $x = \frac{2-y}{3}$ gives x as a function of y. And this equation is equivalent to the original equation $y = 2 - 3x$, since each equation we wrote was justified as being equivalent to the previous equation. This means that the formulas $y = 2 - 3x$ and $x = \frac{2-y}{3}$ are equivalent formulas. This means they are really representing the same relationship between x and y; it's just that one goes from x to y, and the other from y to back to x. In other words, these two equations represent inverse functions!

So if we were given the function $f(x) = 2 - 3x$ and wanted to find a formula for its inverse function, we could just do the calculation above to find that $f^{-1}(y) = \frac{2-y}{3}$.

Pretty simple, huh? But there are a few technical details. You might notice that in the previous section we came up with $f^{-1}(y) = \frac{-(y-2)}{3}$. But take a minute now to convince yourself that these is equivalent expressions. Also, we note that it is usually preferable to switch back to the standard convention of using x for the input into the function, in which case we would write $f^{-1}(x) = \frac{-(x-2)}{3}$.

In the previous homework, we asked you to use tracing paper to sketch another copy of the graph of the function, including the x and y axes, then flip the tracing paper over so that the positive x axis on the tracing now coincides with the positive y axis of the original, so that the positive y axis on the tracing now coincides with the positive x axis of the original. In other words, switch the roles of the x and y variables. You should see that the graph of f will now coincide with the graph of f^{-1}. This demonstrates the one really big idea with inverse functions: that the two functions represent the same relationship between the variables, it's just that the role of input and output has been switched. You can get the graph of the inverse function by just switching the role of x and y in the graph, and similarly, you can find the formula for the inverse function by switching the role of x and y in equation.

Inverse of Exponential and Logarithm Functions. Recall the definition we made for the logarithm:

$$y = \log_b x \text{ means } b^y = x.$$

If we view the equations in the definition of the logarithm as functions, then we can see that the logarithm function is the inverse function for an exponential function, since they result in each other by switching the roles of x and y. That is, if $f(x) = b^x$, for some constant base b, then its inverse function is $f^{-1}(x) = \log_b x$. Similarly, if $g(x) = \log_b x$, then $g^{-1}(x) = b^x$.

Let's illustrate this inverse function relationship by finding a formula for the inverse function for $f(x) = \frac{1}{2}(5 + 3^{x-4})$.

y	$=$	$f(x)$	
y	$=$	$\frac{1}{2}(5 + 3^{x-4})$	Substitute formula for $f(x)$
x	$=$	$\frac{1}{2}(5 + 3^{y-4})$	Switch the x and y. This equation now represents the inverse function $y = f^{-1}(x)$.
$2x$	$=$	$5 + 3^{y-4}$	Multiply both sides by 2
$2x - 5$	$=$	3^{y-4}	Subtract 5 from both sides
$y - 4$	$=$	$\log_3(2x - 5)$	The definition of the logarithm
y	$=$	$\log_3(2x - 5) + 4$	Add 4 to both sides.
$f^{-1}(x)$	$=$	$\log_3(2x - 5) + 4$	Write in function notation

Notice how the original function $f(x) = \frac{1}{2}(5 + 3^{x-4})$ does following sequence of operations on the input x: subtract 4, put as exponent over base 3, add 5, multiply by ½. And now notice how the inverse function $f^{-1}(x) = \log_3(2x - 5) + 4$ does the following sequence of operations: multiply by 2, subtract 5, put into logarithm base 3, add 4. Comparing the two sequences for these inverse functions, we can see that the sequence for the one function undoes everything done by the other function, and in reverse order.

Non-invertible functions. As we saw in the Class Activity, not all functions have an inverse function. If a function has two different inputs that result in the same output, then the inverse process will have one input leading to two different outputs, which violates the definition of a function. Functions that have an inverse function are called **invertible**. In order to be invertible, a function can only have one unique input corresponding to each output. These kinds of functions are called **one-to-one**.

One common function that is not one-to-one, and therefore not invertible, is the squaring function. To explore this, let's see what happens when we apply our technique of "solving for x" to find an inverse function for $f(x) = x^2$. First we can write:

$$y = x^2$$

Now we want to solve for x. Remember there is no property of equality that says you can take the square root of both sides of the equation. What we need to do is think carefully about what it means to "solve the equation $y = x^2$ for x." It means "find the values for x that will make the equation $y = x^2$ true." So the question we have to ask is what number(s) can we square to get y? If y were 25, there'd be two numbers we could square to get y, namely x could be 5 or ⁻5.

$$25 = x^2 \text{ has solutions}$$
$$x = 5 \text{ and } x = {}^-5$$

Note we cannot just say that the solution is $x = \sqrt{25}$. Remember the definition the square root of $\sqrt{25}$ is that it is the non-negative number you square to get 25. So $\sqrt{25}$ is defined to be 5. If we want ⁻5 (and we do, since it's a solution to our equation), and we want to use square roots, we would have to call it $^-\sqrt{25}$. Similarly, if y were 10, there'd be two numbers we could square to get y, namely x could be $\sqrt{10}$ or $^-\sqrt{10}$

$$10 = x^2 \text{ has solutions}$$
$$x = \sqrt{10} \text{ and } x = {}^-\sqrt{10}.$$

In general, there are be two numbers we could square to get y, namely x could be $\sqrt{y}$ or $^-\sqrt{y}$.

$$y = x^2 \text{ has solutions}$$
$$x = \sqrt{y} \text{ and } x = {}^-\sqrt{y}.$$

So if we are applying our solve for x technique to find the inverse function for $f(x) = x^2$, we would get not one function, but two: $f^{-1}(y) = \sqrt{y}$ and $f^{-1}(y) = -\sqrt{y}$. Which one is it? Well neither one by itself satisfies the definition for an inverse function and if we combine them together as a single function, it violates the definition of a function. So we are left with the ugliness of having to say that $f(x) = x^2$ is not invertible.

But we *can* get a unique inverse function if we restrict our domain of $f(x) = x^2$ to consider only either the non-negative case, or the non-positive case. Here's what that would look like. Let's first consider the non-negative case, and define a new function $g(x) = x^2, x \geq 0$. This is just the squaring function, but with a restricted domain that only inputs non-negative numbers. Then if we apply our technique we get

$y = g(x)$ Function notation

$y = x^2$ Substitute the formula for g, **provided $x \geq 0$**

$\sqrt{y} = x$ Definition of the square root, since $x \geq 0$

By restricting the domain of g to be non-negative numbers, we satisfy the definition of the square root, and so the inverse of $g(x) = x^2, x \geq 0$ is $g^{-1}(y) = \sqrt{y}$. Or, if we switch back to x being the input, we get $g^{-1}(x) = \sqrt{x}$. Shown below are graphs of both $g(x)$ and $g^{-1}(x)$.

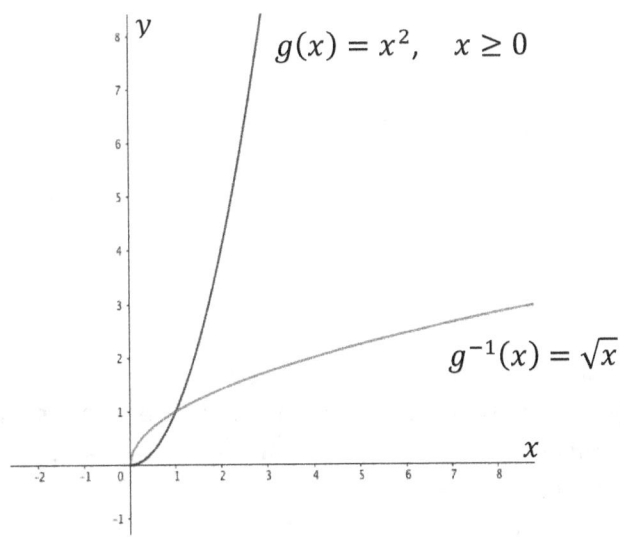

Notice how that graph of $g^{-1}(x)$ has the same shape as the graph of $g(x)$, just is a different orientation. One way to describe this relationship geometrically is that the graphs are reflections of each other across the line $y = x$. Draw the line $y = x$ on the graph above to see this reflection.

Similarly, we could also restrict the domain of the squaring function to just the non-positive values. Here's what that would look like. Let's consider the function $h(x) = x^2$, $x \leq 0$. Then if we apply our technique we get

$$y = h(x) \qquad \text{Function notation}$$

$$y = x^2 \qquad \text{Substitute formula for } h, \textbf{ provided } x \leq 0$$

$$-\sqrt{y} = x \qquad \text{Definition of the square root,}$$
$$\text{Since } \sqrt{y} \geq 0, \textbf{ we have } x \leq 0 \textbf{ as required}$$

By restricting the domain of h to be non-positive numbers, we get the inverse of $h(x) = x^2$, $x \leq 0$ to be $h^{-1}(y) = -\sqrt{y}$. Or, if we switch back to x being the input, we get $h^{-1}(x) = -\sqrt{x}$. Below we show the graphs of $h(x)$ and $h^{-1}(x)$ together. Draw in the line $y = x$ and see how these graphs are reflections of each other across that line.

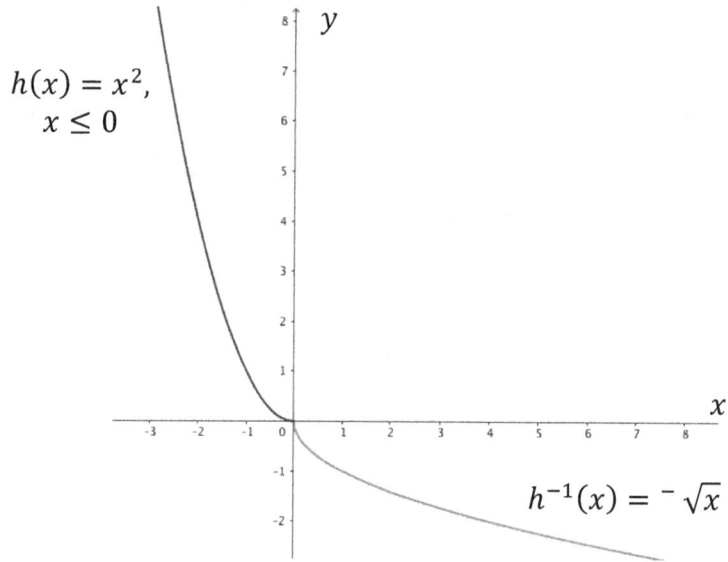

Thinking about inverses for squares and square roots takes some careful thought. But notice how if we always justify our claims by using the definitions, laws and properties, then these justifications will tell us exactly what the domain and range of these functions will need to be.

We will finish this section with another example of how the definition of the square root as being a non-negative number affects finding inverse functions. Consider the function $f(x) = \sqrt{x + 9}$. First we can notice that since this function involves a square root, and since the square root of a negative number is not a real number, the domain of this function is $x \geq -9$. Now, let's apply our technique of solving the function formula for x.

228

$$y = f(x) \qquad \text{Function notation}$$

$$y = \sqrt{x+9} \qquad \text{Substitute the formula for } f$$

$$y^2 = x+9 \qquad \text{Definition of the square root,}$$
provided y is non-negative, that is $y \geq 0$.

$$y^2 - 9 = x + 9 - 9 \qquad \text{Property of Equality: Add } {}^-9 \text{ to both sides}$$

$$y^2 - 9 = x \qquad \text{Since } 9 - 9 = 0$$

So we have that $f^{-1}(y) = y^2 - 9$, provided $y \geq 0$. Or switching to x as the input variable, that $f^{-1}(x) = x^2 - 9$, provided $x \geq 0$. Notice how our solving the equation method gives us not only the formula for the inverse function, but also it tells us its domain.

To summarize, we have shown that if $f(x) = \sqrt{x+9}$, which has a domain of $x \geq {}^-9$, then the inverse function is $f^{-1}(x) = x^2 - 9$, with a restricted domain of $x \geq 0$. Let's sketch the graph of each of these functions. Notice how the graph of $f^{-1}(x) = x^2 - 9$ with the restricted domain $x \geq 0$ means that we only get half of the parabola we would have gotten without the restricted domain.

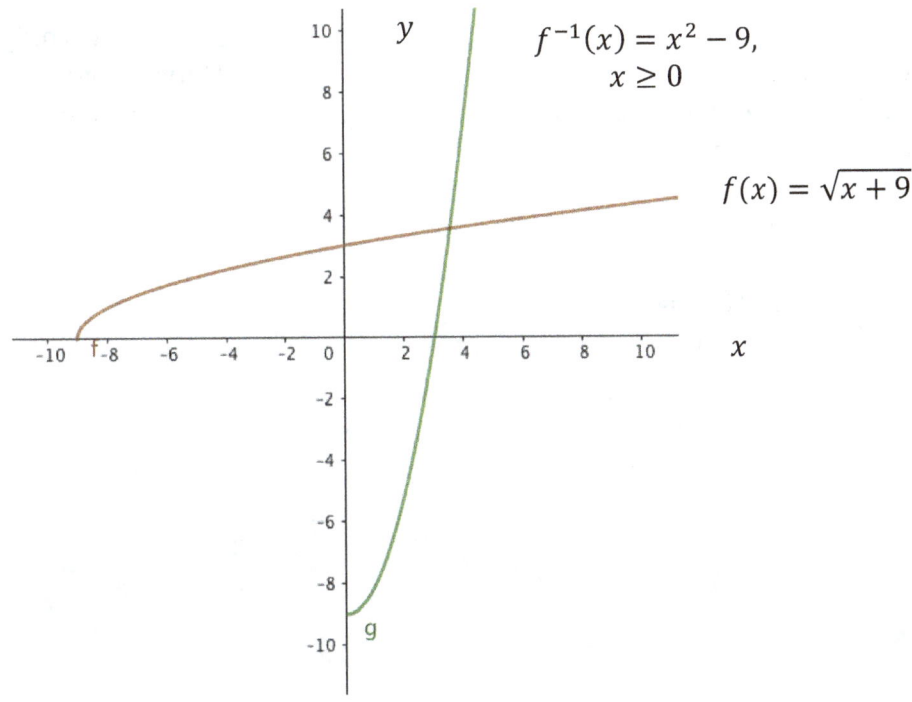

Oh, and go ahead and draw in the line $y = x$ on the graph above to help see how the graphs are reflections of each other.

So we have figured out that the graph of $f(x) = \sqrt{x+9}$ is actually half of a parabola, opening sideways. The other half would be the reflection of this graph over the x-axis, namely, by making all the y values negative. We could define a new function (let's call it h) whose graph would be this other half of the parabola, namely $h(x) = {}^-\sqrt{x+9}$. For fun, let's see what happens if we find the inverse of this function using our "solve for x" technique.

$$y = h(x)$$

$$y = {}^-\sqrt{x+9} \qquad \text{Substitute into function notation}$$

$$y^2 = x+9 \qquad \text{Definition of the square root, } \textbf{provided } y \leq 0.$$

$$y^2 - 9 = x + 9 - 9 \qquad \text{Property of Equality: Add } {}^-9 \text{ to both sides}$$

$$y^2 - 9 = x \qquad \text{Since } 9 - 9 = 0$$

So we can write the inverse function as $h^{-1}(y) = y^2 - 9$. Notice we had to do some careful thinking about the definition of the square root, and this tell us what the domain of this inverse function is. Since the square root $\sqrt{x+9}$ is defined to be a non-negative number, $y = {}^-\sqrt{x+9}$ would have to be a non-positive number, in other words $y \leq 0$.

So we get the inverse function is $h^{-1}(y) = y^2 - 9$, with $y \leq 0$. Or switching to x as the input variable, we can write $h^{-1}(x) = x^2 - 9$, with $x \leq 0$. Notice that it's the same inverse function as before, except with a different domain. We get the other half of the parabola! Here's the graphs:

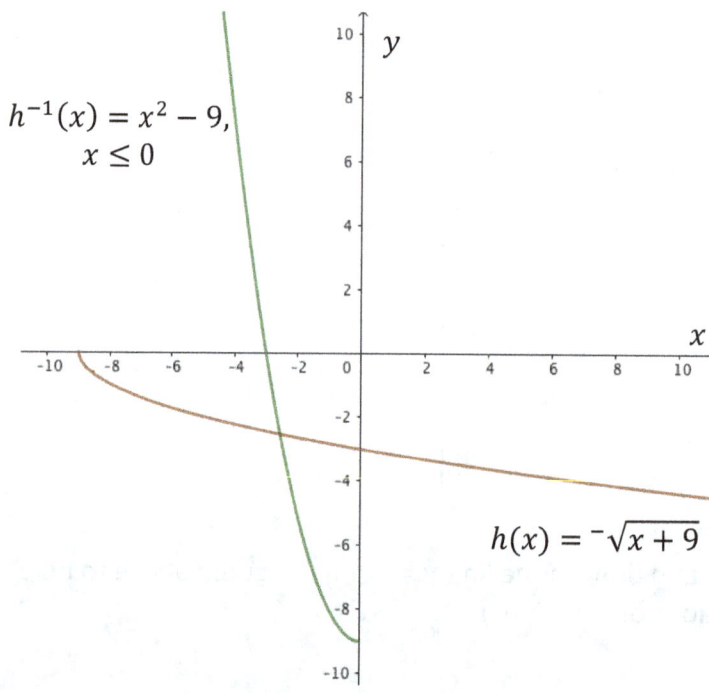

Homework Set 30

1) Sketch the graphs of the two functions we explored at the beginning of the read and study in this section: $f(x) = 2 - 3x$ and $f^{-1}(x) = \frac{^-(x-2)}{3}$, on the same axes. Use tracing paper to demonstrate for yourself that if you flip the graphs so that the x and y axes get switched, that the graphs of f and f^{-1} switch.

2) Find all solutions to the equation $\sqrt{3(x-2)} = 4$.

3) A function $f(x)$ is defined with following sequence of operations:

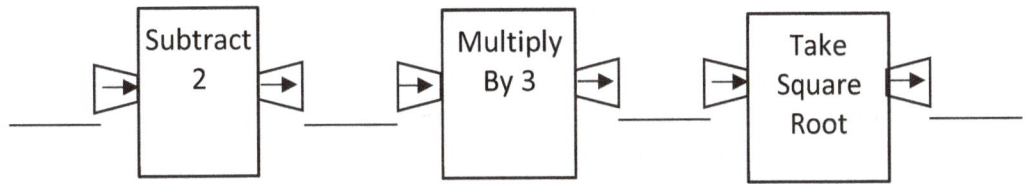

   a. Write a formula for $f(x)$ and state its domain and range.
   b. Sketch an accurate graph of this function for $^-10 < x < 10$ on grid paper
   c. Make a new function machine diagram for the inverse function $f^{-1}(x)$.
   d. Revisit problem 2 above. How do the steps you did in solving that equation compare with the sequence of operations in the inverse function machine?
   e. Sketch an accurate graph of $f^{-1}(x)$ for $^-10 < x < 10$ on grid paper
   f. Use the "Solve for x" technique to find a formula for $f^{-1}(x)$ and state its domain and range.

4) Consider the function defined by the formula $f(x) = \frac{12}{\sqrt{x-2}}$.
   a) Make a machine diagram for f.
   b) State the domain and range for f.
   c) Find $f(3)$
   d) Find $f^{-1}(3)$.
   e) Find a formula for $f^{-1}(x)$.
   f) State the domain and range for f^{-1}.

5) For each function $f(x)$ below, find a formula for its inverse $f^{-1}(x)$
   a) $f(x) = 4 + 5 \cdot e^x$
   b) $f(x) = \frac{2}{3}(2)^{x+1}$
   c) $f(x) = 2\log_3(x+6)$
   d) $f(x) = 3 - \ln(2x)$
   e) $f(x) = \ln(x^2 + 1)$

6) For each function below $f(x)$, find a formula for its inverse $f^{-1}(x)$. State the domain and range of each function and its inverse.
   a) $f(x) = 1 - \frac{4+3x}{5}$.
   b) $f(x) = \frac{x}{1-3x}$.
   c) $f(x) = \sqrt[3]{x+8}$
   d) $f(x) = \frac{4x+2}{3x-6}$.
   e) $f(x) = 3(x+4)^2 - 5$, $x \leq {}^-4$

7) In the Class Activity, we saw that the function $f(x) = 2(x+1)^2$ is not invertible. Now suppose we define a new function with a restricted domain as follows:
$$f(x) = 2(x+1)^2, \qquad x \geq {}^-1.$$

   a) Sketch the graph of this function.
   b) State the domain and range for this function.
   c) Explain now why this new function f is invertible.
   d) Make a machine diagram for the inverse function $f^{-1}(x)$.
   e) Find a formula for the inverse function $f^{-1}(x)$.
   f) Sketch the graph of $f^{-1}(x)$ on the same axes as you showed the graph of $f(x)$.
   g) State the domain and range for $f^{-1}(x)$.

8) Consider the function $f(x) = \sqrt{9 + x^2}$.
   a) Make a function machine diagram that shows this function as a sequence of operations.
   b) Can the formula for this function be simplified to $f(x) = 3 + x$? Make a convincing argument.
   c) The function $f(x) = \sqrt{9 + x^2}$ does not have an inverse (in other words, it's not "invertible"). Why not? Make a convincing argument.
   d) What happens when you apply our solving the formula for x technique?
   e) Create a new function by restricting the domain of $\sqrt{9 + x^2}$ so that this new function has an inverse.
   f) Find the formula for this inverse function and sketch its graph.

9) Could there be a function that is its own inverse? That is, could there be a function where the formula for $f(x)$ is the same as the formula for $f^{-1}(x)$? If not, explain why. If so, find an example and demonstrate that $f(x) = f^{-1}(x)$.

Class Activity 31: Finding the Right Number(s)

Your task is to find exact values for all solutions to the following equations.

1. $2^{3x-1} = 16$

2. $8^x = \left(\dfrac{1}{2}\right)^{x+4}$

3. $3e^{-5x} = 12$

4. $\text{Log}_2(y^2 - 1) = 0$

5. $\log_3\left(\dfrac{x-6}{x+2}\right) = {}^-2$

6. $x + 3 = \sqrt{x^2 + 4}$

7. $(2x-4)^2 = 36$

8. $(x-1)^5 = 35$

9. $\frac{t+1}{t+2} = 3 - \frac{12}{t}$

10. $x + 1 = \sqrt[3]{x^3 + 1}$

11. $x^2 + 3x = (x+3)^2$

12. $\frac{y+6}{3} = \frac{y^2+4}{y}$

Algebra Tiles Template

References:

Give credit where credit is due.

Author unknown (ironically)

- Driscoll, M. (1999) *Fostering Algebraic Thinking: A Guide for Teachers Grades 6 – 10*. Portsmouth, NH: Heinemann.

- *Mathematical Quotations Server* (MQS) at math.furman.edu.

Glossary:

The errors of definitions multiply themselves according as the reckoning proceeds; and lead men into absurdities, which at last they see but cannot avoid, without reckoning anew from the beginning.
  Thomas Hobbes, in J. R. Newman (Ed.) The World of Mathematics

algebraic expression: a combination of symbols involving numbers (numerals, constants and variables) and operations on those numbers (such as addition, subtraction, multiplication, division, powers and roots).

Additive identity: the additive identity is 0, since $a + 0 = a$ and $0 + a = a$ for all numbers.

Additive inverse: the additive inverse of a number a is the number you add to a to get 0. The additive inverse of a is denoted ^-a.

Associative property: A operation * defined on set A satisfies the associative property if $a * (b * c) = (a * b) * c$ for *all* choices of a, b and c in A.

circle: the set of points in the plane that are equidistant from a given point (the center).

Commutative property: A binary operation * defined on set A satisfies the commutative property if $a * b = b * a$ for *all* choices of a and b in A.

complex numbers: This is the set of numbers of the form $a + bi$ where a and b are real numbers and $i = \sqrt{-1}$.

complex conjugate: The conjugate of the complex number $a + bi$ is the number $a - bi$. What's useful about complex conjugates is the fact that the product of a complex number and its conjugate will be a real number.

composition (of functions): A function that is the result of taking the output of one function and putting it into a second function.

conditional (equation): an equation that is true for some value(s) of the variable(s), and false for others.

constant: a fixed, unchanging number.

contradiction (equation): an equation that is never true

239

cubic polynomial: A cubic is a polynomial of degree three. It can be put in the form: $y = ax^3 + bx^2 + cx + d$.

degree (of a polynomial): the largest power of the variable in any term of the polynomial.

dependent variable: In the context of a function, the independent variable (often denoted by x) takes on some domain of values and the dependent variable (often denoted by y) is assumed to vary with respect to changes in x.

discontinuity: a value for x where a function $f(x)$ is not continuous, such as where the function is undefined or has a break, jump or vertical asymptote.

distributive law For all real numbers a, b and c: $a \cdot (b + c) = (a \cdot b) + (a \cdot c)$. The distributive law applies only to multiplication over addition.

division: Division is defined as multiplying by the multiplicative inverse. That is $a \div b = a \cdot b^{-1}$

domain: The domain of a function is the set of values that can be taken on by the independent variable. In other words, the domain is the set of possible inputs for the function.

ellipse: the set of all points in the plane such that the sum of the distances from two given points (the foci) is constant.

end behavior: The end behavior of a function $y = f(x)$ is a description of what happens to the y values as the x values tend to positive or negative infinity.

equation: A statement saying that two expressions have the same numerical value.

equivalent expressions: Two algebraic expressions are equivalent if they have the same value for all relevant values of the variables involved.

equivalent equations: Two equations are equivalent if they have the same set of solutions.

exponential expression: For a positive real number a and real number x, the expression a^x is defined to have the following three properties: $a^0 = 1$, $a^1 = a$, and $a^{x+y} = a^x \cdot a^y$

exponential function: An exponential function is one having the form: $y = a \cdot b^x$ where a and b are parameters.

factor: something that is being multiplied in a product.

function: A function is a relationship between two sets of objects so that each object in the first set (the domain) is paired with exactly one object in the second set (the range).

function notation: $y = f(x)$ means that y in the range is the output when x from the domain is the input into the function called f.

functional thinking is relating a value of a variable to the value of another variable

fundamental theorem of algebra: Every polynomial of degree n has exactly n complex roots (counting multiple roots), and can be written as the product of n linear factors.

graph of an equation: the set of all ordered pairs that make the equation true.

horizontal asymptote: a function $y = f(x)$ has a horizontal asymptote at $y = a$ if the y values approach a as x tends to positive or negative infinity.

identity (equation): an identity is an equation that is true for all possible values for the variable(s) we are interested in.

identity (element): The identity element with respect to an operation * is an element e in A such that $a * e = a$ and $e * a = a$ for *all* choices of a in A.

independent variable: In the context of a mathematical function, the independent variable (often denoted by x) takes on some domain of values and the dependent variable (often denoted by y) is assumed to vary with respect to changes in x.

integers: The set of integers is the set: {...-3, -2, -1, 0, 1, 2, 3 ...}. This is the set of all natural numbers together with their additive inverses and zero.

inverse (elements): An element x is the *inverse* of an element y with respect to an operation * if $x * y = e$ and $y * x = e$, where e is the identity element for the operation.

inverse functions: Two functions f and g are inverses if $f(g(x)) = x$ and $g(f(x)) = x$.

irrational number: A real number that *cannot* be written as the fraction of two integers. These numbers have decimal names that neither terminate nor repeat.

linear expression: A linear expression is one in which each variable is of degree one.

linear function: A degree one polynomial, often in the form: $y = mx + b$ where m and b are parameters.

logarithm function: The function $y = \log_b x$ is the inverse function for the exponential function $x = b^y$.

modeling: Modeling is finding a mathematical representation (algebraic, numeric or graphic model) which captures the essential elements of a situation.

multiplicative identity: the multiplicative identity is 1, since $a \cdot 1 = a$ and $1 \cdot a = a$ for all numbers a.

multiplicative inverse: the multiplicative inverse of a number a is the number you multiply to a to get 1. The multiplicative inverse of a is denoted a^{-1}.

natural numbers: The natural numbers is the set of "counting numbers": {1, 2, 3, 4, 5, ... }.

nonlinear expression: A nonlinear expression is one in which one of the variable has degree other than one (such as quadratic, square root, or rational expressions).

n^{th} root: the nth root of a number is the number you raise to the nth power to get that number. If there is only one such number, it can be positive or negative. But if there are two such numbers, then the nth root is taken to be the non-negative one.

numeral: a symbol to represent a specific constant number.

parabola: the set of all points in the plane that are equidistant from a given point (the focus) and a given line (the directrix).

parameter: A parameter is an unspecified constant in a mathematical expression, equation, or function.

polynomial: A function that can be put in the form:

$$y = a_n x^n + a_{n-1} x^{n-1} + \ldots + a_2 x^2 + a_1 x + a_0 \text{ with } a_n \neq 0.$$

property of equality: Adding or subtracting the same number, or multiplying or dividing the same non-zero number to both sides of an equation does not change the solutions to the equation.

quadratic function: A quadratic function is a polynomial of degree two. Two common forms are vertex form: $y = a(x - h)^2 + k$ and standard form: $y = ax^2 + bx + c$ (where $a, h,$ and k or $a, b,$ and c are parameters).

rational function: a function that is the fraction (ratio) of two polynomials.

rational number: a number that can be written as a fraction (ratio) of two integers. These numbers have decimal names that either terminate or repeat.

range: The range of a function is the set of values that can be taken on by the dependent variable. In other words, the range is the set of possible outputs of the function.

real numbers: This is the set of all possible distances and their additive inverses. This set can be modeled geometrically by the set of points on a line (often called the *real number line*).

recursive thinking: relating a value of a variable to its previous value(s)

root (of a function): A root of a function $y = f(x)$ is value for x that makes $y = 0$. On a graph, a root is an x-value where the function touches or crosses the x-axis.

root (of a number): The nth **root of x** is the number you raise to the nth power to get x.

slope of a line: The slope of a line is a constant value defined as the ratio of the change in y to the change in x. This value is often denoted by m in the function $y = mx + b$.

solution: A solution to an equation is a value variable or variables that makes the equation true.

solving an equation: Finding all the values for the variable or variables that makes the equation true.

square root: the square root of a number is the non-negative number you square to get that number. That is, $\sqrt{x} = y$ means that y is the non-negative number such that $(y)^2 = x$.

subtraction: Subtraction is defined as adding the additive inverse. That is $a - b = a + {}^-b$

term: something that is being added in a sum.

variable a symbol which represents an unknown quantity or a changing quantity.

vertex: the point on a graph (such as an ellipse or parabola) where the curve turns to go back the opposite direction.

zero product property: If a product is zero, then one or both of the factors must be zero.

www.ingramcontent.com/pod-product-compliance
Lightning Source LLC
Chambersburg PA
CBHW080540220526
45466CB00010B/2978